Springer Theses

Recognizing Outstanding Ph.D. Research

Aims and Scope

The series "Springer Theses" brings together a selection of the very best Ph.D. theses from around the world and across the physical sciences. Nominated and endorsed by two recognized specialists, each published volume has been selected for its scientific excellence and the high impact of its contents for the pertinent field of research. For greater accessibility to non-specialists, the published versions include an extended introduction, as well as a foreword by the student's supervisor explaining the special relevance of the work for the field. As a whole, the series will provide a valuable resource both for newcomers to the research fields described, and for other scientists seeking detailed background information on special questions. Finally, it provides an accredited documentation of the valuable contributions made by today's younger generation of scientists.

Theses are accepted into the series by invited nomination only and must fulfill all of the following criteria

- They must be written in good English.
- The topic should fall within the confines of Chemistry, Physics, Earth Sciences, Engineering and related interdisciplinary fields such as Materials, Nanoscience, Chemical Engineering, Complex Systems and Biophysics.
- The work reported in the thesis must represent a significant scientific advance.
- If the thesis includes previously published material, permission to reproduce this must be gained from the respective copyright holder.
- They must have been examined and passed during the 12 months prior to nomination.
- Each thesis should include a foreword by the supervisor outlining the significance of its content.
- The theses should have a clearly defined structure including an introduction accessible to scientists not expert in that particular field.

More information about this series at http://www.springer.com/series/8790

Antonio Fernández-Ruiz

Extracellular Potentials in the Hippocampus

Sources and Biophysical Mechanisms

Doctoral Thesis accepted by
the Complutense University of Madrid, Spain

Springer

Author
Dr. Antonio Fernández-Ruiz
Department of Applied Physics III,
Faculty of Physics
Complutense University,
Ciudad Universitaria
Madrid
Spain

Supervisors
Prof. Miguel Sancho Ruiz
Department of Applied Physics III,
Faculty of Physics
Complutense University,
Ciudad Universitaria
Madrid
Spain

Prof. Sagrario Muñoz San Martín
Department of Applied Physics III,
Faculty of Physics
Complutense University,
Ciudad Universitaria
Madrid
Spain

ISSN 2190-5053 ISSN 2190-5061 (electronic)
Springer Theses
ISBN 978-3-319-41038-8 ISBN 978-3-319-41039-5 (eBook)
DOI 10.1007/978-3-319-41039-5

Library of Congress Control Number: 2016942520

Printed on acid-free paper

This Springer imprint is published by Springer Nature
The registered company is Springer International Publishing AG Switzerland

A ti que te fuiste antes de tiempo
pero nos dejaste lo mejor que tenemos:
tu recuerdo, tu fuerza y tu alegría.

Tu hijo

Supervisors' Foreword

The thesis focuses on the study of the generation and propagation of local field potentials (LFPs) in the hippocampus, taking rodents as the experimental model. It presents an innovative combination of signal analysis techniques with computational modelling of LFP propagation to shed light on the origin of extracellular potentials and the dynamics of the hippocampal circuitry. The current sources are obtained from high/density two-dimensional recordings by the inverse Current Source Density method (CSD) and the resulting spatiotemporal information about the current dynamics is decomposed into functional components using Independent Component Analysis (ICA). Then, a finite element modeling (FEM) of the field propagation is developed, showing the often neglected importance of the macroscopic anatomical structure on the recorded LFPs.

The coordination of the hippocampal network with the entorhinal cortex (EC) input is investigated by analyzing theta-gamma cross-frequency coupling. Three sources of gamma oscillations in the CA1 region with different distribution, spectral characteristics, and theta–phase preference are identified. It is demonstrated that, during different behavioral states and stages of a memory task, there are differences in the relative contributions of CA3 and EC inputs to CA1 theta-modulated gamma activity and spiking. These inputs can cooperate or compete to modulate CA1 output and interregional communication. The thesis provides both detailed information about methodology and background on the field and important insights into the entorhinal–hippocampal network.

Madrid, Spain
May 2016

Prof. Miguel Sancho Ruiz
Prof. Sagrario Muñoz San Martín

Abstract

During recent decades, the study of the nervous system function has moved gradually from the cellular and molecular levels to a more systemic perspective. Due to the recent developments in technologies to record and manipulate large populations of cells in behaving animals, the available volume of data has rapidly increased. Although this boost in the amount of data regarding the function of neural circuits in action opens new and exciting roads to understanding information processing in the brain, it also poses new challenges to their analysis and interpretation. One variable that can capture the activity of multiple neural populations in different brain areas with high spatial and temporal resolution is the local field potential (LFP). Despite the potential usefulness of LFPs to read out the computations performed by brain circuits during behavior, its study has been hampered by the difficulties in extracting meaningful information from it. In the present thesis, we aim to study the biophysical basis of the LFP as a way to reach a deeper understanding of it in terms of its underlying physiological mechanisms.

We focused on the rodent hippocampus due to the important role that it plays in many cognitive functions (such as memory, learning, and spatial navigation) and the advantages of its simplified layered structure for the study of LFPs. Taking advantage of high-density silicon probe recordings in behaving rats, we were able to map the spatiotemporal distribution of LFPs along the dorsoventral and transversal axes of the hippocampus. We implemented a novel method consisting in the combination of Independent Component Analysis (ICA) and Current Source Density (CSD) analysis to separate and identify the synaptic sources of hippocampal LFPs. In addition, we built a tridimensional model of the rat dorsal hippocampus where the spatiotemporal characteristics of those sources were implemented and LFPs were simulated by means of the finite elements method. Those simulations reproduced the common hippocampal LFP patterns and laminar characteristics, thus enabling us to reach a better understanding of the generation of LFPs in the hippocampus and verifying the accuracy of the ICA + CSD decomposition of experimental data.

To prove the relevance of the knowledge gained about the biophysical basis of LFPs in the hippocampus, we addressed the issue of theta and gamma dynamics in the hippocampal-entorhinal circuit during different behavioral states. We found three sources of gamma oscillations in the CA1 region with different laminar distribution, spectral characteristics, and theta-phase preference. Slow gamma (30–60 Hz), elicited by the CA3 input to the CA1 pyramidal cells, peak on the descending theta phase; while mid-frequency gamma (60–120 Hz), brought about by entorhinal layer three input to CA1, dominate on the peak of the theta cycle. A third source of very fast oscillations (100–180 Hz) peaked on the theta trough and was of local origin. Those theta-coupled gamma oscillations were present in both exploration and REM sleep albeit with different strengths. We showed that during different behavioral states and stages of a memory task, CA3 and entorhinal gamma inputs could compete or cooperate to modulate CA1 output and interregional communication.

The results presented in this thesis constitute an advance in our understanding and interpretation of LFPs and brain oscillations. They highlight the importance of the employment of adequate experimental and analytical methods to investigate the activity of brain circuits and point to the LFP as a useful although complex variable in this purpose.

Publications List of the Author

Schomburg EW*, **Fernández-Ruiz A***, Berényi A, Mizuseki K, Anastassiou CA, Koch C, Buzsáki G. (2014) *Theta phase segregation of input-specific gamma patterns in entorhinal-hippocampal networks.* Neuron. 84:470–485. (*co-first authors).

Benito N*, **Fernandez-Ruiz A***, Makarov VA, Makarova J, Korovaichuk A, Herreras O. (2014) *Spatial blocks of coherent pathway-specific LFPs in the hippocampus reflect different modes of presynaptic synchronization.* Cereb Cortex. 24:1738–52. (*co-first authors).

Enriquez-Barreto L, Cuesto G, Dominguez-Iturza N, Gavilán E, Ruano D, Sandi C, **Fernández-Ruiz A**, Martín-Vázquez G, Herreras O, Morales M. (2014) *Learning improvement after PI3K activation correlates with de novo formation of functional small spines.* Front Mol Neurosci 2;6:54.

Fernández-Ruiz A*, Schomburg EW*. (2013) *The rules of entrainment: are CA1 gamma oscillations externally imposed or locally governed?* J Neurosci 33:19045–19047. (*co-first authors).

Fernandez-Ruiz A, Muñoz S, Sancho M, Makarov VA, Herreras O. (2013) *Cytoarchitectonic and dynamic origins of giant positive LFPs in the Dentate Gyrus.* J Neurosci 33:15518–15532.

Fernández-Ruiz A, Herreras O. (2013) *Identifying the synaptic origin of ongoing neuronal oscillations through spatial discrimination of electric fields.* Front Comput Neurosci 7:5.

Fernandez-Ruiz A, Makarov VA, Herreras O. (2012) *Sustained increase of spontaneous input and spike transfer in the CA3-CA1 following long-term potentiation in vivo.* Front Neural Circuits 6:71.

Fernandez-Ruiz A, Makarov VA, Benito N, Herreras O. (2012) *Schaffer-specific local field potentials reflect discrete excitatory events at gamma-frequency that may fire postsynaptic hippocampal CA1 units.* J Neurosci 32:5165–5176.

Acknowledgements

Since I was an undergrad student in Sevilla starting to discover what Science was really about and, in particular, after finishing my studies and initiating my research career, many people have helped me to arrive to the point where I am now.

Starting from those early times in the School of Biology in the University of Sevilla and later in the Complutense of Madrid, I have to remember a few of my professors who had the patience and energy to show me their research, let me be around them, and answer my thousand questions: Josefina Cano, Ángel Pastor, Sara Morcuende, Esperanza Rodriguez, Carmen Rúa and many others. Even if I was convinced to try to become a scientist, I was not yet sure that I would choose the neurosciences. I learned to enjoy many different fields of biology, such as evolution, genetics, and zoology, thanks especially to the amazing written works of great scientists (e.g., SJ Gould, Jacob and Monod, Darwin, and so many others) and of some enthusiastic professors. Sadly, not all my professors or classmates were equally enthusiastic about science and research; however, with some of them who truly were, I shared many interesting and stimulating talks. Among those, I remember and mention Alvarito, Gonzalo, Miguel, Curro, Alberto, and David.

After finishing my biology studies, my interest in the physical sciences grew more and more, plus I felt the necessity of complementing my formation with some basic formal knowledge about physics and maths. I would probably have never developed such interest if it were not for reading the inspiring writing of some of the brightest minds of the past century: Heisenberg, Schrodinger, Einstein, etc. I really enjoyed those readings as well as learning about quantum mechanics, particle physics and logic. However, I realized that was not the field where I could do my best work, so, despite those temptations, I finally found my place in the field of neuroscience.

During my Physics Masters, I met great and helpful professors (Miguel Sancho, Sagrario Muñoz, Tatiana Alieva, Fernando Sols, and others) and mates from the Schools of Physics and Mathematics (Pilar, Pablo, Ricardo, David, Jaime, and Jesus). And I also met a very special person.

Although I had some previous research experience I really started doing science in the Cajal Institute. The years I spent there constitute in general a good experience. In the laboratory of Óscar, I learned many things: to love electrophysiology, to perform experiments, to plan and carry out my own research project, etc. I was more than lucky with some of my lab mates, Azahara and Gonzalo, with whom I shared the enthusiasm of discovery and the long hours that it usually takes. The time I shared with them both inside and outside the lab was the best part of all that I spent in Madrid. I was also very lucky with some of the other students of the Institute for so many good moments and conversations, especially Maria, Jorge and Edu; and some of the researchers, Liset, Gonzalo. However, during those years, I also got to know the worst face of academic life and the conflicts that arise when people are more driven by other motivations than the pursuit of knowledge and the enjoyment of Science both as a career and as a lifestyle.

After my stage in the Cajal, I went to the laboratory of Gyuri Buzsáki in NYU. That was the best of the luck and a step forward into the real international academic world, beyond the idiosyncratic borders of the Spanish scientific community. I am very grateful to the generosity, energy, and enthusiasm of Gyuri, who embraced me and showed me how far I could go once I found the proper path. I am also grateful to many people who generously shared their data or codes with me, and others that helped me with my experiments: Erik, Tony, Kenji, Dion, Yuta, Gabrielle, and others. I have to give special thanks to Tony Berényi for sharing his great data and advice and to help me to do crazy silicon probe implantations, both in NYU and, after that, in his laboratory in the University of Szeged. Also to my Hungarian mates Misi, Ancsa, Gábor and Gergö.

Finally I came back to accomplish my Ph.D. at the Complutense. I have the greatest gratitude to Miguel and Sagrario who, despite the complicate situation I brought to them, had no hesitation to embrace me and help with all possible things; personal, scientific and administrative. Without that the present thesis would have not been possible.

Although the funding climate and support for research was far from ideal in Spain in recent years, I was lucky to have had, during my university and research stages, the economic support of different institutions: Universidad de Sevilla, the Xlab and Georg-August Universität of Gőttingen, Universidad Complutense de Madrid, Ministerios de Educación y de Ciencia, CSIC and JAE, European Molecular Biology Organization (EMBO), Fundación La Caixa, Sociedad Española de Neurociencia (SENC), and Federation of European Neuroscience (FENS).

The Ph.D. work was many times rewarding and enjoyable, but not always. In those times, I found one of my greatest shelters and sources of renewed energy in books. The words of some people with very similar concerns, worries, or hopes to mine often helped me during this time: Borges, Hesse, Dostoyevski, Wittgenstein, Cernuda, Machado, Asimov, and many many others.

Finally, many people from outside the academic world have helped me to arrive to this point, in particular, my family. I have to double thank Azahara: on the personal side, for being always at my side with her constant love, understanding and (not so easy to grant so often) patience, from one corner of the world to the other; and on the professional side for helping me with experiments and analyses regardless if it was a holiday or late at night. To my brother for not losing his trust in me and never giving up despite to my forgetfulness and self-absorbed temper and for his constant support. To my father for his endless interest in everything I do and for teaching me the value of hard work, persistence, and self-improvement since I was a child. And to my mother, for her infinite love, happiness, and optimism. For teaching me the best way to face any adversity, to never lose faith in myself, or give up and for showing me that the future is always full of opportunities for those who are willing to find them.

Contents

Acronyms

CA1/3	*Cornus Ammoni* 1/3
CFC	Cross Frequency Coupling
CSD	Current Source Density
CWT	Complex Wavelet Transform
DG	Dentate Gyrus
DNQX	6,7-dinitroquinoxaline-2,3-dione
EC	Entorhinal Cortex
EEG	Electroencephalogram
ESPS	Excitatory post-synaptic potential
FEM	Finite Elements Method
GABA	γ-aminobutyric acid
GC	Granular cell
IC	Independent Component
ICA	Independent Component Analysis
IPSP	Inhibitory post-synaptic potential
LEC	Lateral entorhinal cortex
LFP	Local Field Potential
LPP	Lateral Perforant Path
MEC	Medial entorhinal cortex
MI	Modulationindex
MPP	Medial Perforant Path
NMDA	*N*-Methyl-D-aspartic acid or *N*-Methyl-D-aspartate
SEM	Standard Error of the Mean

Chapter 1
Introduction

1.1 Genesis and Study of the Macroscopic Electrical Activity in the Brain

The brain is a complex system, probably the most complex of all. Despite that, it is formed by relatively simple (but only in comparative terms) elements: neurons and other cell types. Its complexity mainly arises from the myriad of intricate ways those elements are interconnected forming networks or neural circuits. The dominant paradigm in systems neuroscience in last decades has been that the distributed activity of brain circuits is the physiological mechanism underlying cognitive functions [10, 17, 75, 79]. There is also a large degree of functional modularity in the brain, that is, discrete brain regions dedicated to particular functions, in particular in early sensory systems. However, most cognitive functions rely on the activity of "higher" cortical associative areas which integrate information from different sensory modalities together with emotional and motivational content. In addition to the integration of external inputs to conform unified percepts, another complementary function of these associative brain areas is to support the interaction of external inputs with self-generated internal ones and stored memory engrams [18, 56].

To perform those functions, brain circuits need to be able to integrate distributed local processes or computations into globally organized states and, in turn, to route the flow of highly processed information to downstream structures both in the brain and effector organs (muscles, etc.). For this wide integration and segregation of information, brain oscillations, or "rhythms", are thought to perform an essential role [13, 23, 57, 79].

Brain oscillations are present across all animal phyla, from invertebrates to birds, reptiles and mammals [23, 61]. Although their mechanisms vary largely, a common feature is the requirement of synchronized activity in specific circuits of interconnected excitatory and inhibitory neurons which generate rhythmic postsynaptic potentials (PSPs). Those PSPs synchronized over large populations of cells are the

A. Fernández-Ruiz, *Extracellular Potentials in the Hippocampus*,
Springer Theses, DOI 10.1007/978-3-319-41039-5_1

main source of macroscopic oscillations. These rhythmic synaptic inputs produce in the target neural population alternate windows of enhanced and reduced excitability. This mechanism plays a double function: on the one hand it favor the integration of incoming inputs and their plasticity in discrete time frames, while, on the other, it segregates the output of the neurons (the generation of synchronized action potentials) into short burst and sequences [14, 82].

Neural oscillations span several orders of magnitude in frequency, from the very slow (<1 Hz) to the very fast (200–500 Hz; Fig. 1.1a). Among the most studied of them are the slow-wave-sleep delta waves (0.5–1.5 Hz), the hippocampal theta rhythm (5–10 Hz), the cortical gamma rhythms (40–90 Hz) and hippocampal ripples (140–220 Hz). Slow oscillations are coherent across wide regions even spanning different structures, generating synchronized membrane fluctuations in widespread neuronal networks. On the other hand, faster oscillations are usually associated with local computations performed by small numbers of cells and thus are only synchronous in a highly restricted volume.

Oscillations of different frequencies very often interact with each other both locally and across structures. This interaction follows typically a hierarchical fashion, thus the phase of the slower oscillations modulates the amplitude and occurrence of the faster ones ([9, 26, 27, 35, 72]. Fig. 1.1b, c). Cross-frequency coupling has been shown to correlate with memory performance in both humans and animals [3, 26, 71,76]. Cross-frequency coupling in cortical circuits has been proposed as a mechanism to integrate local computations across modules in a broader time scale, thus allowing the efficient integration of different streams of information necessary for most cognitive processes [57].

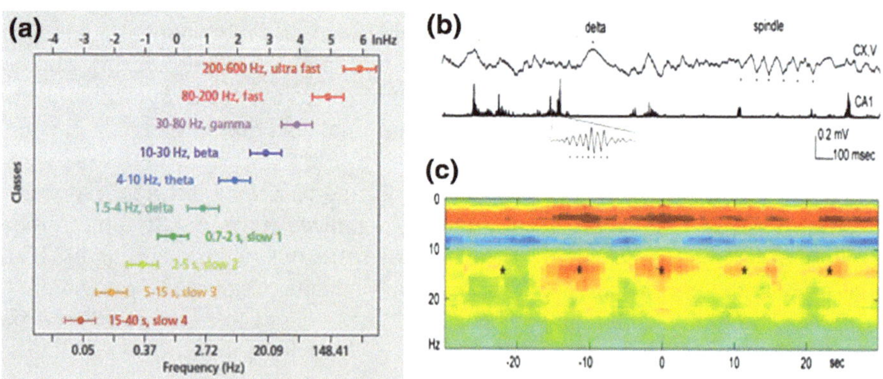

Fig. 1.1 **a** Main classes of brain oscillations spanning several orders of magnitude. **b** LFP trace from rat cortical layer 5 during sleep showing different characteristic oscillations: delta waves and spindles. Below, filtered and rectified LFP from hippocampus CA1 pyramidal layer displaying ripples as large amplitude bursts. **c** Ripple-triggered power spectrogram of cortical LFP showing modulation by spindles (revealed by a power increase of approximately 15 Hz). Both events are also modulated by the slow oscillation (0–3 Hz) [23]

The interest in studying brain oscillations arises also from the fact that particular rhythms are altered during different pathological states, including but not limited to epilepsy, Alzheimer's disease, schizophrenia and depression [68, 77]. Thus, its study may lead to a better understanding of the mechanisms of those diseases or even provide early markers for their detection.

To understand the neurophysiological basis of cognitive processes, there are many different approximations. The most traditional approach has been the recording of individual neuronal activity to correlate their firing dynamics with specific behaviors. With modern recording techniques, it is possible to simultaneously record extracellularly a few hundred neurons or to identify intracellularly single cells in behaving animals. On the other extreme of the scale, we find functional imaging techniques (fMRI, MEG, PET, etc.) which can monitor whole brain activity and identify regions that are activated during a particular task. These techniques have also the advantage that they are non-invasive, so they can be used in humans. Despite the advances in recent decades on both extremes, the subcellular and global scales, there is still an important gap in the middle: how individual cells coordinate in neural circuits to orchestrate different cognitive functions. This has prevented us from reaching a comprehensive explanation of behavior and cognition in terms of their underlying physiological mechanisms. The study of neuronal circuit dynamics tries to bridge this gap.

One technique that allows the recording of neural activity of intermingled neural populations in different brain areas with high spatial and temporal resolution is the recording of the intracranial electroencephalogram (EEG). The EEG is produced by the superposition in the extracellular medium of transmembrane currents in adjacent neurons. These currents are elicited mainly by the plethora of synaptic inputs generating PSPs, but there are also other non-synaptic sources of the EEG (see [22]). The transmembrane currents elicit an electric potential that varies dynamically in time and space, giving rise to the recorded EEG (Fig. 1.2). Thus, the EEG contains all the summed activity of the multiple synaptic inputs and the local activity in the region where the extracellular electrode is located. For historical reasons and despite the physical inaccuracy of the term, the intracranially recorded EEG is always referred in the literature as the local field potential (LFP).

Biophysical principles underlying LFP generation have long been known [22, 33, 60]. However, there are still many theoretical and technical problems that limit the usefulness of LFPs to understand brain function. A long-standing biophysical problem that has remained stubbornly within the field for decades is the difficulty to identify the synaptic sources of LFPs so as to correlate activity of known neuronal populations with ongoing behavior. As the LFP in any point in the brain is produced by the activity of intermingled cellular populations with overlapped synaptic territories, to identify the cellular origin of particular LFP patterns, the so-called *inverse problem*, is a complex task. This problem can be also reformulated as follows: once given an experimental macroscopic signal, the amplitude of which varies at different sites (i.e., the LFP), how can the location and extension of the generating source be determined? Multiple combinations of independent sources (transmembrane currents in discrete dendritic domains of a population of

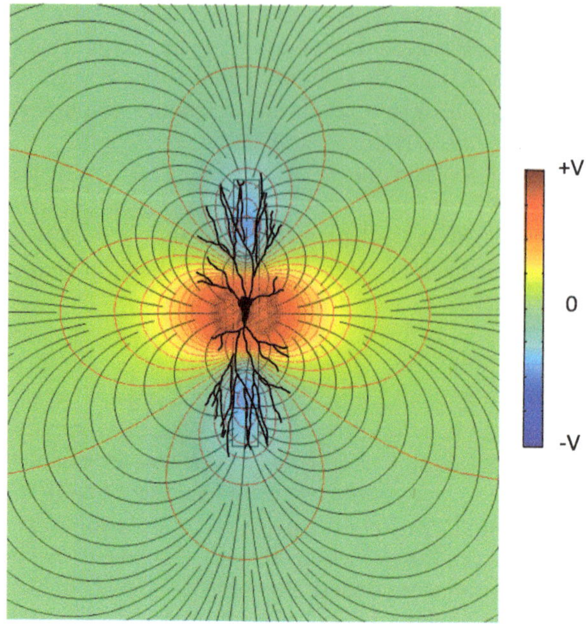

Fig. 1.2 A simplified simulation of a single neuron receiving an inhibitory somatic input. Active outward currents in the soma are compensated by return inward current in the apical and basal dendrites resulting in a negative-positive-negative extracellular potential profile. Colors indicate extracellular voltage, *red lines* indicate isopotential surfaces, and *black* current lines

synchronously active neurons) may give rise to a recorded signal with the same spatial pattern. There is no unique solution, and in most cases, it is difficult or impossible to confirm the potential solutions experimentally.

A common feature of electric fields in the brain is that they vary spatially in a complex manner, on account of the shifting activation of neuron generators with irregular morphology and distribution. Neurons with dominant axial geometry act as strong current dipoles [22, 33, 40, 60] and as such are the main contributors to field potentials. A common afferent input to one or another subcellular domain of the neuron population will give rise to different extracellular potential distributions. This fact underlies the characteristic laminar profile of LFPs in layered structures. The problem arises when several inputs are co-activated, as is usually the case. In such circumstances, the electric currents mix unevenly at different sites, and field potential distributions become complex and variable. Thus, only high-density recordings simultaneously performed at several positions can correctly map for spatial variations of LFPs. Multisite linear recordings are well-suited to a method that has been employed to find the current generators underlying field potentials, known as current source density (CSD) analysis [42]. This approach has been very useful to determine the contributing cells and the location of synaptic membranes activated by afferent stimuli in laminar structures, such as the hippocampus or neocortex [9, 54]. However, while interpreting CSD maps is simple for voltage profiles elicited by stimulating only one afferent pathway, their application to ongoing LFPs renders complex spatial maps of intermingled inward and outward currents, and in general it is not feasible to identify the multiple synaptic generators from them [36] (Fig. 1.3).

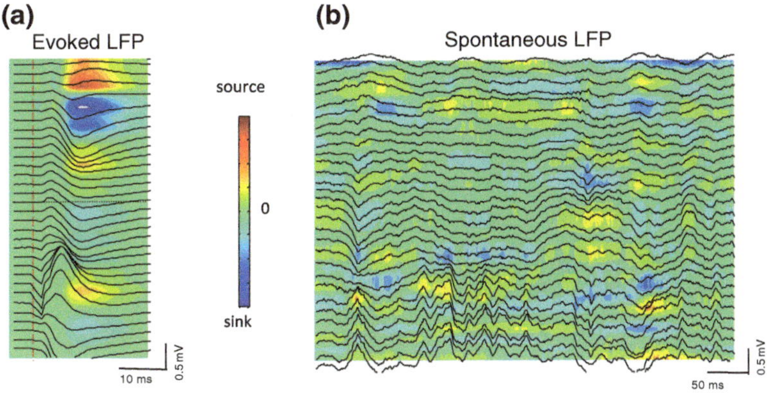

Fig. 1.3 **a** LFP profile and CSD map of an evoked potential in the hippocampus resulting from the stimulation of the CA3. In the upper part (CA1), a strong sink (in *blue*) results from the depolarization of CA1 pyramidal cell dendrites and is compensated by two sources elicited by passive return currents. The recurrent collaterals of CA3 axons also produce a local current dipole. **b** During spontaneous activity in the hippocampus of an awake rat, multiple synaptic inputs arrive simultaneously to the CA1 and CA3 regions making it very complex to interpret CSD distributions [36]

An important additional source of complexity and confounding factors, when analyzing or modeling LFPs or other macroscopic variables of neuronal activity, are the electric properties of brain tissue. For the sake of simplicity, it is commonly assumed that the brain tissue is homogeneous, isotropic, and has ohmic (linear) properties. The contribution to the recorded LFP from a particular neuronal membrane domain (either in a single cell or in a population of synchronously active neurons) is usually estimated as follows

$$\phi(r_e, t) = \frac{1}{4\pi\sigma} \sum_{n=1}^{N} \frac{I_n(t)}{|\mathbf{r_e} - \mathbf{r_n}|} \tag{1.1}$$

The above equation denotes that the contribution of the transmembrane current I_n in a membrane domain n to the LFP recorded at $\boldsymbol{r}_e$ is inversely proportional to the distance between the electrode and the source and to the extracellular conductivity of the tissue σ. The LFP thus would be the result of the linear summation of all the transmembrane currents weighted by their distances to the recording electrode (Fig. 1.4). Here, we are assuming a quasistatic approximation of Maxwell equations in which the electric and magnetic fields are decoupled, and there are not inductive effects. This seems reasonable given the characteristic low frequency of brain activity and in particular of that contributing to the generation of LFPs, mostly below 300 Hz [66, 69]. The above equation also implies that the medium where

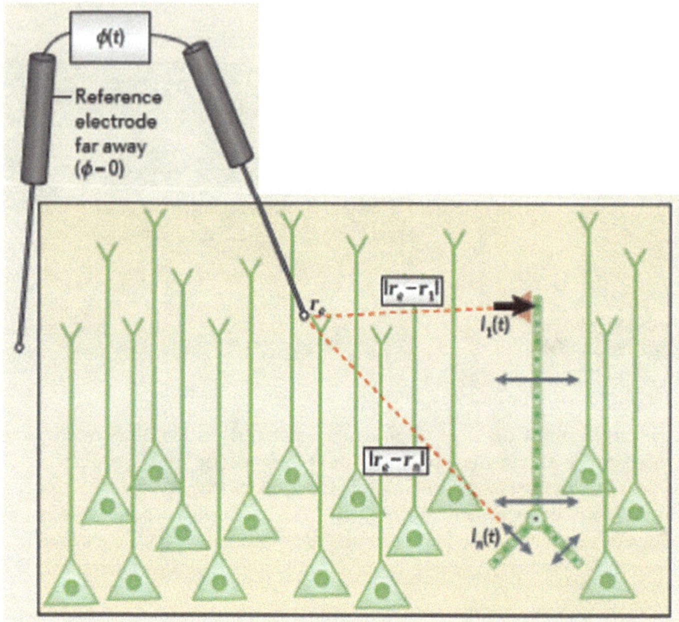

Fig. 1.4 Simplified schematic showing the calculation of the LFP produced by a single pyramidal neuron receiving an apical input. Synaptic input elicits inward transmembrane currents locally ($I_1(t)$, *black arrow*) which are followed by return outward currents all along the membrane of the cell ($I_n(t)$, *grey arrows*). Transmembrane currents in each of the cell compartments elicit an electric potential at the tip of the recording electrode (r_e) with respect to a distant reference point. The summation of all the transmembrane currents in each compartment n, weighted by their distance to the electrode $|\boldsymbol{r}_e - \boldsymbol{r}_n|$ and the conductivity of the tissue, is the LFP produced by that neuron. Modified from [33]

neuronal currents propagate is an infinite, homogeneous, and isotropic conducting volume. Although in many cases those assumptions may be reasonable, they are fundamentally inadequate. On a large scale, the brain is obviously not homogeneous, for example, the presence of the liquid filled ventricles may greatly distort the propagation of electric fields. At a microscopic level, the tissue can be considered mainly homogeneous, but even if so, it has been shown that it is not completely isotropic due to the presence of dense cellular layers or myelinated axonal bundles [66]. A third assumption made in the above equation is that the conductivity is purely ohmic, that is, it has no frequency-dependent or capacitive effects. About this last point, there has been much debate in recent years, and there is not yet a definite answer [4, 33, 58]. It can be said that there are probably some frequency-dependent properties of brain tissue, but those are not very influential in most cases and applications.

1.2 Anatomy and Physiology of the Hippocampal Formation

1.2.1 Anatomical and Functional Organization of the Hippocampal Formation

Over the decades, the mammalian hippocampal formation (HF) (Fig. 1.5a) has attracted the attention of neuroscience researchers because of its easily recognizable anatomical features [25, 59], its prominent electrophysiological activity patterns [20, 81], and its functional implication in learning and memory processes. For comprehensive overview reading of hippocampal formation, anatomy, and function, there are many excellent reviews [7, 32, 80, 83] and books [2, 13]. Here we provide a brief summary of the most relevant aspects for the present thesis.

The term hippocampal formation (HP) is used to designate several brain structures derived from the medial pallium and strongly related, both anatomically and functionally: the hippocampus proper or *cornusammonis*, the Dentate Gyrus (DG), and the subiculum [11].

In the hippocampus, cellular bodies are densely packed to form one layer. Pyramidal cells form the *cornus ammonis*, divided by Lorente de Nó into four sub-regions CA1–4. In the DG, granular cells form the U-shaped granular layer [25, 59]. The subiculum has several cellular layers and is divided into three main regions: presubiculum, parasubiculum, and postsubiculum. The HF is strongly interconnected with the adjacent entorhinal cortex (EC), which constitutes the

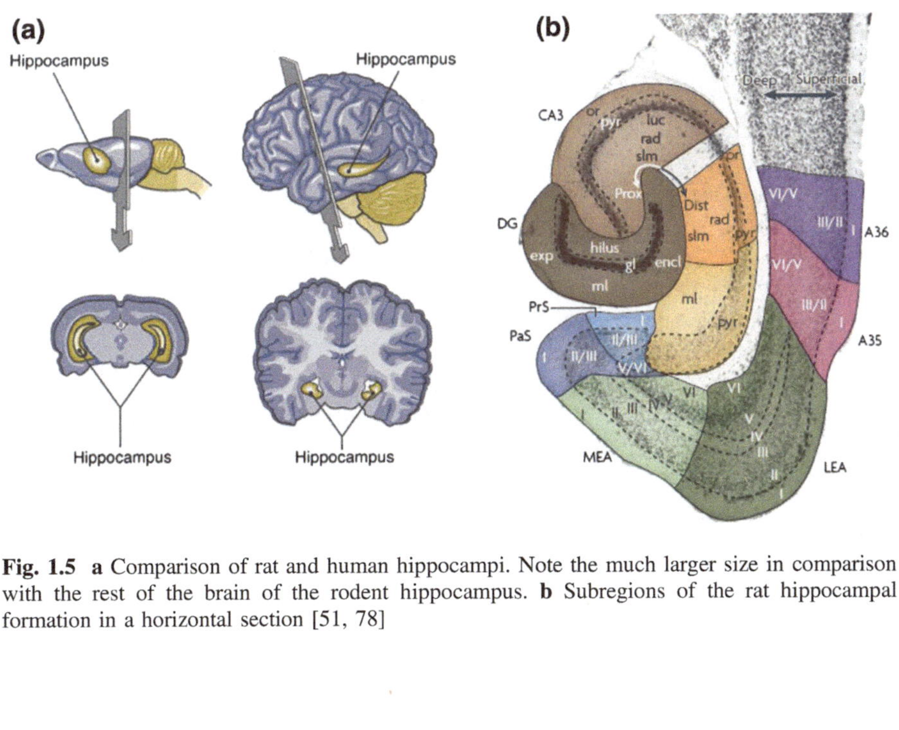

Fig. 1.5 **a** Comparison of rat and human hippocampi. Note the much larger size in comparison with the rest of the brain of the rodent hippocampus. **b** Subregions of the rat hippocampal formation in a horizontal section [51, 78]

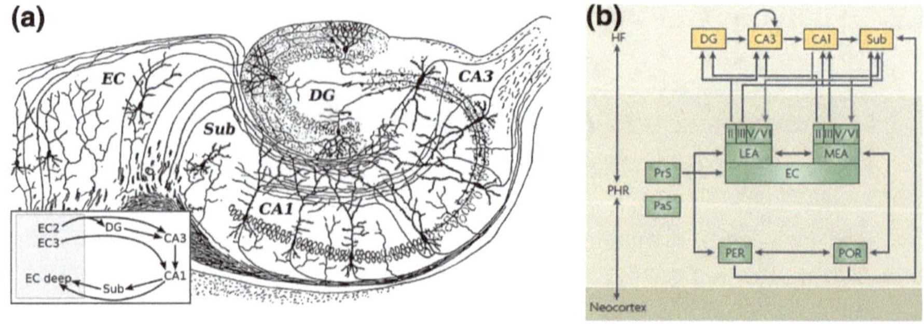

Fig. 1.6 **a** Modified from an original drawing of Cajal, showing the hippocampal subdivisions with its main cellular types and axonal pathways. **b** Diagram of the classic schema of information flow in the hippocampal formation. Modified from [24, 78]

principal source of afferences to it. It possesses a more complex cytoarchitectonic structure than the hippocampus with six layers. It is divided into two functionally and anatomically distinct regions, the medial (MEC) and the lateral portions (LEC; Fig. 1.5b).

Classically, the hippocampal formation has been described as a feedforward circuit of excitatory synapses (Fig. 1.6a). According to this model, neurons in the layer II (L2) of the EC project to the DG through the perforant pathway. L2 axons also innervate CA3 pyramidal cells. DG granular cells send their axons (the so called mossy fibers) to CA3. CA3 pyramidal cells axons divide to make recurrent connections within CA3 and also innervate a large extent of the CA1 region (the Schaffer pathway). CA1 pyramidal cells also receive direct connections from EC L3 cells through the temporo-ammonic pathway, and in turn, project to the subiculum and the deep layers of the entorhinal cortex (Fig. 1.6b).

In addition to these excitatory connections, the activity in every node of the network is modulated by inhibitory synapses made by the large existing number and types of interneurons. In the HP, as well as in most of the other regions of the brain, and together with the excitatory principal cells, there are multiple classes of those inhibitory cells (Fig. 1.7). The different types of interneurons have very different intrinsic properties, firing patterns, and targets. Thus, different roles and involvement in network activity have been proposed for them [43, 53].

One important characteristic of the HP and a big advantage to the study of their LFPs, as we will see below, is the stratification of their inputs. Different afferences to all hippocampal subregions have well defined dendritic domains. As it will be the focus of subsequent work, we will cover the case of DG and CA1. In the DG, MEC L2 axons establish synapses in the middle portion of the dendritic arbor of the granular cells while LEC L2 axons arrive at its outer part (Fig. 1.8a). The inner portion of the dendritic arbor is innervated by commissural and associational fibers. CA1 pyramidal cells have both basal and apical dendrites. The basal dendrites form the stratum oriens and the much larger apical ones extend into two sublayers: stratum radiatum and stratum lacunosum-moleculare. CA3 axons innervate the

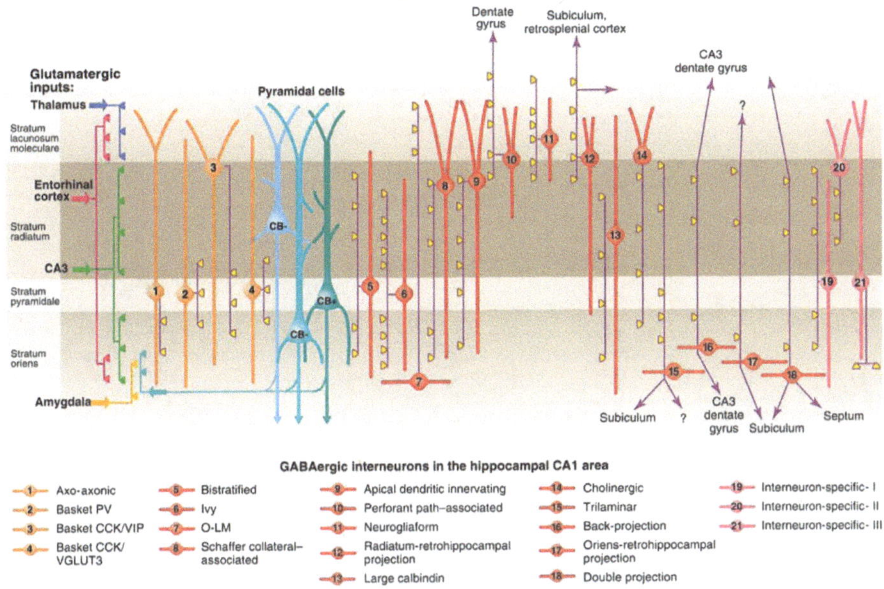

Fig. 1.7 Subtypes of interneurons in the CA1 region. More than 20 types of interneurons have been defined in the hippocampus. Note the variability in synaptic domains of different interneurons onto CA1 pyramidal cells [53]

stratum radiatum and stratum oriens while entorhinal fibers arrive at the stratum lacunosum-moleculare and, to lesser extent, also at the stratum oriens (Fig. 1.8b). Thalamic axons are also circumscribed to the stratum lacunosum-moleculare. It is worth to mention that both excitatory inputs also innervate several interneuron types, which in turn project synapses onto the principal cells and other interneurons.

1.2.2 Hippocampal LFP Patterns

The hippocampus displays very prominent LFPs that are strongly correlated with the behavior of the animal, thus were early used to study hippocampal function [81]. During active locomotion and exploration and also during rapid-eye movement (REM), sleep hippocampal LFPs are dominated by a strong quasi-sinusoidal oscillation with a narrow power spectrum peaking around seven to nine Hz, known as theta oscillations (see Fig. 1.9a) [20]. Theta oscillations are coordinated in all hippocampal subregions and along the whole extent of the hippocampus [21, 67]. They are also present, at the same time and with high coherence, in many other regions of the brain, mainly in the limbic system: entorhinal cortex, subiculum, piriform cortex, septum, etc. [1]. Over recent decades, there has been intense research to elucidate the generating mechanism of theta oscillations but the debate

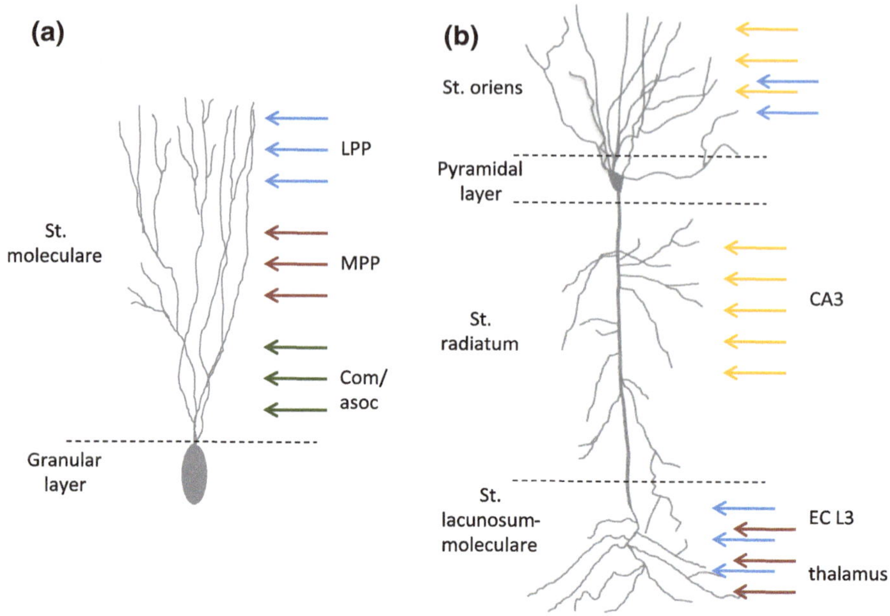

Fig. 1.8 Diagram depicting the different sub-layers of the DG (**a**) and CA1 (**b**) regions of the hippocampus and their main afferences

is still open (see [12] for a comprehensive review). There is a consensus that the main theta rhythm generator is in the medial septum-diagonal band of Broca, which imposes its rhythm onto the other regions. However, some of these regions in isolation, even in vitro, can generate their own theta oscillations [45]. When it comes to the generator of theta LFPs in the hippocampus and other regions, the picture is even more complex. Almost every hippocampal sublayer has its own theta current generator that possesses current dipoles which contribute to the overall theta LFP [12, 55]. During different behaviors, those theta dipoles are coordinated in a variable manner [64] modulating accordingly the spiking of hippocampal neurons. Pharmacological manipulations have been proved useful to dissociate and characterize the different synaptic generators of the theta rhythm in the hippocampus and other areas [6, 65, 74], but we are still far from a comprehensive understanding of its mechanisms.

A rather opposite electrographic state to the one just described is present during immobility, consummatory behaviors, and slow-wave sleep (SWS), and it is characterized by more asynchronous activity and the presence of sharp-wave ripples (SWR) complexes. SWR are LFP patterns composed by a large negative "sharp wave" in the stratum radiatum of CA1 accompanied by very fast (120–180 Hz) oscillations or "ripples" in the CA1 pyramidal layer (see Fig. 1.9b) [9, 38, 62]. The CA1 sharp-wave is produced by a large depolarizing conductance in the apical dendrites of the pyramidal cells elicited by a highly synchronous firing of a large

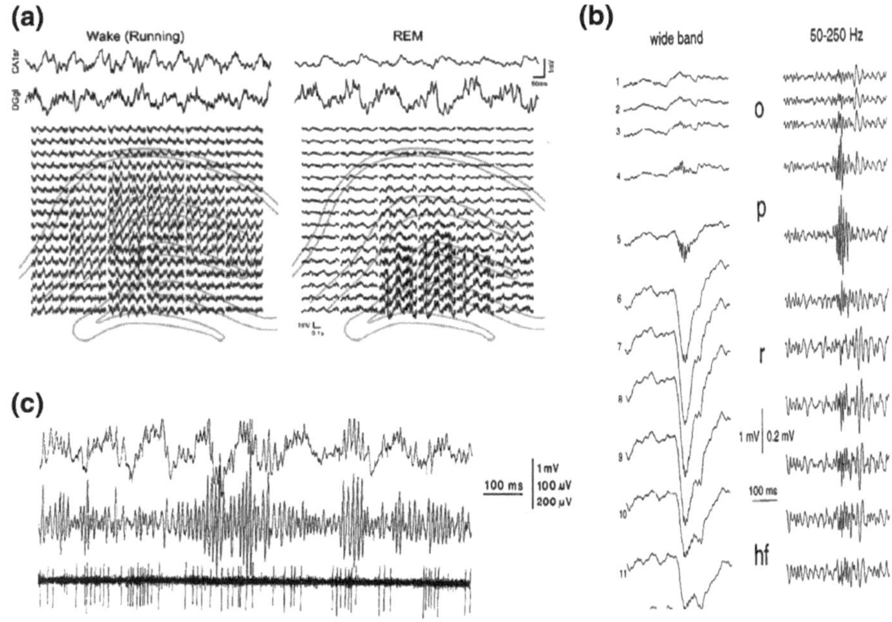

Fig. 1.9 **a** Theta oscillations in the hippocampus during active running and REM sleep appear across every subregion. Note the different power and frequency of oscillations in both behavioral states. **b** SWR complexes are characterized by a strong negative deflection in the stratum radiatum (r) and simultaneous fast oscillation in the CA1 pyramidal layer (p). **c** Gamma oscillations appear in the hippocampus in every layer, are modulated by the concomitant theta rhythm, and entrain the firing of numerous local neurons. (Modified from [8, 63, 84]

population of CA3 pyramidals. It has been proposed that its generation is controlled by the CA3 recurrent network and perisomatic-targeting interneurons [46]. The generation of CA1 ripples is less understood, and several mechanisms ranging from inhibitory fast post-synaptic potentials (IPSPs), action potentials or a combination have been proposed [52, 70, 84]. What is clear is that a large population of both CA1 pyramidal cells and interneurons are firing synchronously during SWR [29], thus the output of the hippocampus to its target regions is enhanced. Much of the interest about SWR comes from the fact that they are strongly related to learning and memory consolidation (see [15] for a comprehensive review).

Another ubiquitous LFP pattern that is not only in the hippocampus but in almost every brain region, mainly in the cortex is that of gamma oscillations. Gamma oscillations (30–90 Hz) were originally studied in the neocortex and are related to different cognitive functions such as attention, sensory integration, and learning [44, 57, 73]. Several mechanisms have been proposed to explain its emergence and coordination [19, 30, 34, 37, 46]. Most of them imply the interplay between excitation and inhibition in local networks. Of particular importance is the role of fast-spiking, perisomatic-targeting interneurons that are thought to synchronize principal cells spiking on the gamma time scale to form cell-assemblies [47, 53].

In the hippocampus, gamma oscillations are present concomitantly with other oscillations, mainly theta, whose phase modulates gamma amplitude (see Fig. 1.9c) [9, 27, 28]. Although initially gamma oscillations were considered a unitary phenomenon, evidence suggests that there is an enormous diversity of mechanisms and functions implying different gamma patterns. In every hippocampal subregion, gamma oscillations of different frequency, ranging from the slow gamma spectrum (30–50 Hz) to the very fast (100–200 Hz), can be observed and have different laminar distributions, synaptic mechanisms, and behavioral correlates [5, 9, 30, 38, 39, 41, 71, 85].

At the core of the neural computations performed by the hippocampus is the synchronous activation of neuronal assemblies whose temporal dynamics may govern the processing and flow of information in brain circuits. According to the "cell assembly" hypothesis [48, 50], information in the brain is represented by groups of synchronously firing neurons, whose participation reflects an interaction between sensory input and internally generated patterns. A prominent role in forming these assemblies is ascribed to gamma oscillations. For example, place cells representing the same spatial position fire together in the time window of gamma cycles and are often phase-locked to the same gamma frequency [49]. Neuronal assemblies organize in specific temporal sequences, which have been shown to encode past (recall) and future (planning) aspects of the behavior of the animal [31]. A postulated mechanism for generating assembly sequences is the interaction among the multitude of brain oscillations organized by cross-frequency coupling [16]. In the hippocampus and entorhinal cortex, the phase of theta rhythm has been shown to modulate the power of gamma oscillations according to behavioral demands, as well as the firing of both principal cells and interneurons. Thus, the function of theta rhythm may be to organize on a broader scale the gamma time-scale cell assemblies [57].

1.3 Goals of the Thesis

The general aim of the present thesis is to perform an experimental and computational study of the biophysical and physiological basis of macroscopic brain signals. For this purpose, we will focus on the rodent hippocampal LFPs.

The first goal of the thesis will be a methodological one. We will implement a novel approach to analyze and interpret the large amount of information contained in large-scale recordings of LFPs in behaving rodents. In particular, we will tackle the problem of separating and identifying the contributing synaptic sources of the LFPs (the "inverse problem"). As a first step, this goal will be addressed by the application of advanced mathematical tools such as Independent Component Analysis (ICA) and Current Source Density (CSD) analysis. For an effective use of these methods, it would be necessary to make a detailed spatial mapping of LFPs along different axis of the structure of interest, the hippocampus. As a second step, we will build a tridimensional model of the rat dorsal hippocampus. In such a model, the spatiotemporal characteristics of hippocampal LFP sources will be implemented and the corresponding electric potential distributions solved for the whole structure by means of finite elements method (FEM), the "forward problem"). The comparison of the results from the experiments and simulations will enable us to reach a better understanding of the generation of LFPs in the hippocampus and of the accuracy of the solutions found for their inverse problem.

The second goal will be to characterize the oscillatory dynamics of hippocampal LFP sources during different behavioral states. We will perform a time-frequency decomposition of the different LFP sources with a special emphasis on the generation of gamma oscillations and their coordination by the theta rhythm. This analysis will be performed during sleep, exploration, and learning to study if theta-gamma dynamics in the hippocampus can shed light on the mechanisms of information processing by hippocampal circuits during behavior.

We expect that the successful achievement of the aforementioned goals will contribute towards a deeper understanding of the generation of macroscopic brain signals and will lead us to realize a better interpretation of neural oscillations in terms of their underlying physiological mechanisms.

References

1. Alonso A, García-Austt E (1987) Neuronal sources of theta rhythm in the entorhinal cortex of the rat. I. Laminar distribution of theta field potentials. Exp Brain Res 67:493–501
2. Andersen P, Morris RG, Amaral D, Bliss T, O'Keefe J (2007) The hippocampus book. Oxford University Press, New York
3. Axmacher N, Henseler MM, Jensen O, Weinreich I, Elger CE, Fell J (2010) Cross- frequency coupling supports multi-item working memory in the human hippocampus. Proc Natl Acad Sci USA 107:3228–3233
4. Bédard C, Destexhe A (2011) Generalized theory for current-source-density analysis in brain tissue. Phys Rev E 84:041909
5. Belluscio MA, Mizuseki K, Schmidt R, Kempter R, Buzsáki G (2012) Cross-frequency phase-phase coupling between θ and γ oscillations in the hippocampus. J Neurosci 32:423–435
6. Benito N, Fernandez-Ruiz A, Makarov VA, Makarova J, Korovaichuk A, Herreras O (2014) Spatial modules of coherent activity in pathway-specific LFPs in the hippocampus reflect topology and different modes of presynaptic synchronization. Cereb Cortex 24:1738–1752
7. Bland BH, Oddie SD (2001) Theta band oscillation and synchrony in the hippocampal formation and associated structures: the case for its role in sensorimotor integration. Behav Brain Res 127:119–136
8. Bragin A, Jandó G, Nádasdy Z, Hetke J, Wise K, Buzsáki G (1995a) Gamma (40–100 Hz) oscillation in the hippocampus of the behaving rat. J Neurosci 15:47–60
9. Bragin A, Jandó G, Nádasdy Z, van Landeghem M, Buzsáki G. (1995b) Dentate EEG spikes and associated interneuronal population bursts in the hippocampal hilar region of the rat. J Neurophysiol 73:1691–1705
10. Bullmore E, Sporns O (2009) Complex brain networks: graph theoretical analysis of structural and functional systems. Nat Rev Neurosci 10(3):186–198
11. Burwell RD, Witter MP, Amaral DG (1995) Perirhinal and postrhinal cortices of the rat: a review of the neuroanatomical literature and comparison with findings from the monkey brain. Hippocampus 5:390–408

12. Buzsáki G (2002) Theta oscillations in the hippocampus. Neuron 33:325–340
13. Buzsáki G (2006) Rhythms of the brain. Oxford University Press
14. Buzsáki G (2010) Neural syntax: cell assemblies, synapsembles, and readers. Neuron 68: 362–385
15. Buzsáki G (2015) Hippocampal sharp wave-ripple: a cognitive biomarker for episodic memory and planning. Hippocampus [Epub ahead of print]
16. Buzsáki G, Draguhn A (2004) Neuronal oscillations in cortical networks. Science 304: 1926–1929
17. Buzsáki G, Mizuseki K (2014) The log-dynamic brain: how skewed distributions affect network operations. Nat Rev Neurosci 15:264–278
18. Buzsáki G, Moser EI (2013) Memory, navigation and theta rhythm in the hippocampal-entorhinal system. Nature Publishing Group 16:130–138
19. Buzsáki G, Wang X-J (2012) Mechanisms of Gamma Oscillations. Annu Rev Neurosci
20. Buzsáki G, Leung LS, Vanderwolf CH (1983) Cellular bases of hippocampal EEG in the behaving rat. Brain Res 287:139–171
21. Buzsáki G, Czopf J, Kondakor I, Kellenyi L (1986) Laminar distribution of hippocampal rhythmic slow activity (RSA) in the behaving rat: current source density analysis, effects of urethane and atropine. Brain Res 365:125–137
22. Buzsáki G, Anastassiou CA, Koch C (2012) The origin of extracellular fields and currents—EEG, ECoG, LFP and spikes. Nat Rev Neurosci 13:407–420
23. Buzsáki G, Logothetis N, Singer W (2013) Scaling brain size, keeping timing: evolutionary preservation of brain rhythms. Neuron 80:751–764
24. Cajal S (1901) Estudios sobre la corteza cerebral humana. IV estructura de la corteza cerebralolfativa del hombre y mamíferos. Trabajos del Lab Inv Biol Univ Madrid 1:1–140
25. Cajal S (1911) Histologie du Systemenerveux de l'homme et des vertebres, vol II. Maloine, Paris
26. Canolty RT, Edwards E, Dalal SS, Soltani M, Nagarajan SS, Kirsch HE, Berger MS, Barbaro NM, Knight RT (2006) High gamma power is phase-locked to theta oscillations in human neocortex. Science 313:1626–1628
27. Chrobak JJ, Buzsáki G (1998) Gamma oscillations in the entorhinal cortex of the freely behaving rat. J Neurosci 18:388–398
28. Colgin LL, Denninger T, Fyhn M, Hafting T, Bonnevie T, Jensen O, Moser M-B, Moser EI (2009) Frequency of gamma oscillations routes flow of information in the hippocampus. Nature 462:353–357
29. Csicsvari J, Hirase H, Mamiya A, Buzsáki G (2000) Ensemble patterns of hippocampal CA3-CA1 neurons during sharp wave-associated population events. Neuron 28:585–594
30. Csicsvari J, Jamieson B, Wise KD, Buzsáki G (2003) Mechanisms of gamma oscillations in the hippocampus of the behaving rat. Neuron 37:311–322
31. Dragoi G, Buzsáki G (2006) Temporal encoding of place sequences by hippocampal cell assemblies. Neuron 50:145–157
32. Eichenbaum H (2004) Hippocampus: cognitive processes and neural representations that underlie declarative memory. Neuron 44:109–120
33. Einevoll GT, Kayser C, Logothetis NK, Panzeri S (2013) Modelling and analysis of local field potentials for studying the function of cortical circuits. Nat Rev Neurosci 14:770–785
34. Engel AK, Fries P, Singer W (2001) Dynamicpreditions: oscillation and synchrony in top-down processing. Nat Rev Neurosci 2:704–716
35. Fell J, Axmacher N (2011) The role of phase synchronization in memory processes. Nat Rev Neurosci 12:105–118
36. Fernández-Ruiz A, Herreras O (2013) Identifying the synaptic origin of ongoing neuronal oscillations through spatial discrimination of electric fields. Front Comput Neurosci 7:5
37. Fernández-Ruiz A, Schomburg EW (2013) The rules of entrainment: are CA1 gamma oscillations externally imposed or locally governed? J Neurosci 33:19045–19047
38. Fernández-Ruiz A, Makarov VA, Benito N, Herreras O (2012a) Schaffer-specific local field potentials reflect discrete excitatory events at gamma frequency that may fire postsynaptic hippocampal CA1 units. J Neurosci 32:5165–5176

39. Fernández-Ruiz A, Makarov VA, Herreras O (2012b) Sustained increase of spontaneous input and spike transfer in the CA3-CA1 pathway following long-term potentiation in vivo. Front Neural Circ 6:71
40. Fernández-Ruiz A, Muñoz S, Sancho M, Makarova J, Makarov VA, Herreras O (2013) Cytoarchitectonic and dynamic origins of giant positive local field potentials in the dentate gyrus. J Neurosci 33:15518–15532
41. Fisahn A, Pike FG, Buhl EH, Paulsen O (1998) Cholinergicinduction of network oscillations at 40 Hz in thehippocampus in vitro. Nature 394:186–189
42. Freeman JA, Nicholson C (1975) Experimental optimization of current source-density technique for anuran cerebellum. J Neurophysiol 38:369–382
43. Freund TF, Buzsáki G (1996) Interneurons of the hippocampus. Hippocampus 6:347–470
44. Fries P, Nikolic D, Singer W (2007) The gamma cycle. Trends Neurosci 30:309–316
45. Goutagny R, Jackson J, Williams S (2009) Self-generated theta oscillations in the hippocampus. Nat Neurosci 12:1491–1493
46. Hájos N, Paulsen O (2009) Network mechanisms of gamma oscillations in the CA3 region of the hippocampus. Neural Netw 22:1113–1119
47. Hájos N, Pálhalmi J, Mann EO, Németh B, Paulsen O, Freund TF (2004) Spike timing of distinct types of GABAergic interneuron during hippocampal gamma oscillations in vitro. J Neurosci 24:9127–9137
48. Harris KD (2005) Neural signatures of cell assembly organization. Nat Rev Neurosci 6: 399–407
49. Harris KD, Csicsvari J, Hirase H, Dragoi G, Buzsáki G (2003) Organization of cell assemblies in the hippocampus. Nature 424:552–556
50. Hebb DO (1949) The organization of behaviour
51. Hiller-Sturmhöfel S, Swartzwelder HS (2004) Alcohol's effects on the adolescent brain—what can be learned from animal models (http://pubs.niaaa.nih.gov/publications/arh284/213-221.htm)
52. Ibarz JM, Foffani G, Cid E, Inostroza M, de la Prida LM (2010) Emergent dynamics of fast ripples in the epileptic hippocampus. J Neuroscience 30:16249–16261
53. Klausberger T, Somogyi P (2008) Neuronal diversity and temporal dynamics: the unity of hippocampal circuit operations. Science 321:53–57
54. Leung LS (1979) Potentials evoked by alvear tract in hippocampal CA 1 region in rats. II. Spatial field analysis. J Neurophysiol. 42:1571–1589
55. Leung LS (1984) Model of gradual phase shift of theta rhythm in the rat. J Neurophysiol 52:1051–1065
56. Lisman JE, Idiart MA (1995) Storage of 7 +/− 2 short-term memories in oscillatory subcycles. Science 267:1512–1515
57. Lisman JE, Jensen O (2013) The theta-gamma neural code. Neuron 77:1002–1016
58. Logothetis NK, Kayser C, Oeltermann A (2007) In vivo measurement of cortical impedance spectrum in monkeys: implications for signal propagation. Neuron 55:809–823
59. Lorente de Nó R (1934) Studies on the structure of the cerebral cortex. II. Continuation of the study of the ammonic system. J Psychol Neurol 46:113–177
60. Lorente de Nó R (1947) Analysis of the distribution of action currents of nerves in volume conductors. In: A study of nerve physiology, part 2, vol 132. The Rockefeller Institute, New York, pp 384–477
61. Marder E, Bucher D (2007) Understanding circuit dynamics using the stomatogastric nervous system of lobsters and crabs. Annu Rev Physiol 69:291–316
62. Montgomery SM, Buzsàki G (2007) Gamma oscillations dynamically couple hippocampal CA3 and CA1 regions during memory task performance. Proc Natl Acad Sci USA 104: 14495–14500
63. Montgomery SM, Sirota A, Buzsáki G (2008) Theta and gamma coordination of hippocampal networks during waking and rapid eye movement sleep. J Neurosci 28:6731–6741
64. Montgomery SM, Betancur MI, Buzsáki G (2009) Behavior-dependent coordination of multiple theta dipoles in the hippocampus. J Neurosci 29:1381–1394

65. Newman EL, Gillet SN, Climer JR, Hasselmo ME (2013) Cholinergic blockade reduces theta-gamma phase amplitude coupling and speed modulation of theta frequency consistent with behavioral effects on encoding. J Neurosci, 33:19635–46
66. Nunez PL, Srinivasan R (2006) Electric fields of the brain: the neurophysics of EEG, 2nd edn. Oxford University Press, New York
67. Patel J, Fujisawa S, Berényi A, Royer S, Buzsáki G (2012) Traveling theta waves along the entire septotemporal axis of the hippocampus. Neuron 75:410–417
68. Pittman-Polletta BR, Kocsis B, Vijayan S, Whittington MA, Kopell NJ (2015) Brain rhythms connect impaired inhibition to altered cognition in schizophrenia. Biol Psychiatry 77: 997–1116
69. Plonsey R, Heppner DB (1967) Considerations of quasi-stationarity in electrophysiological systems. Bull Math Biophys 29:657–664
70. Schomburg EW, Anastassiou CA, Buzsáki G, Koch C (2012) The spiking component of oscillatory extracellular potentials in the rat hippocampus. J Neurosci 32:11798–11811
71. Schomburg EW, Fernández-Ruiz A, Berényi A, Mizuseki K, Anastassiou CA, Koch C, Buzsáki G (2014) Theta phase segregation of input-specific gamma patterns in entorhinal-hippocampal networks. Neuron 84:470–485
72. Schroeder CE, Lakatos P (2009) Low-frequency neuronal oscillations as instruments of sensory selection. TINS 32:9–18
73. Singer W, Gray CM (1995) Visual feature integration and the temporal correlation hypothesis. Annu Rev Neurosci 18:555–586
74. Soltesz I, Deschénes M (1993) Low- and high-frequency membrane potential oscillations during theta activity in CA1 and CA3 pyramidal neurons of the rat hippocampus under ketamine-xylazine anesthesia. J Neurophysiol 70:97–116
75. Tononi G, Edelman GM, Sporns O (1998) Complexity and coherency: integrating information in the brain. Trends Cogn Sci 2:474–484
76. Tort ABL, Komorowski RW, Manns JR, Kopell NJ, Eichenbaum H (2009) Theta-gamma coupling increases during the learning of item-context associations. Proc Natl Acad Sci USA 106:20942–20947
77. Uhlhaas PJ, Singer W (2006) Neural synchrony in brain disorders: relevance for cognitive dysfunctions and pathophysiology. Neuron 52:155–168
78. vanStrien NM, Cappaert NLM, Witter MP (2009) The anatomy of memory an interactive overview of the parahippocampal-hippocampal network. Nat Rev Neurosci 10:272–282
79. Varela F, Lachaux JP, Rodriguez E, Martinerie J (2001) The brainweb: phase synchronization and large-scale integration. Nat Rev Neurosci 2:229–239
80. Vinogradova OS (2001) Hippocampus as comparator: role of the two input and two output systems of the hippocampus in selection and registration of information. Hippocampus 11:278–598
81. Whishaw IQ, Vanderwolf CH (1973) Hippocampal EEG and behavior: changes in amplitude and frequency of RSA (theta rhythm) associated with spontaneous and learned movement patterns in rats and cats. BehavBiol 8:461–484
82. Wilson MA, McNaughton BL (1994) Reactivation of hippocampal ensemble memories during sleep. Science 265:676–679
83. Witter MP, Groenewegen HJ, da Silva FHL, Lohman AH (1989) Functional organization of the extrinsic and intrinsic circuitry of the parahippocampal region. Prog Neurobiol 33:161–253
84. Ylinen A, Bragin A, Nádasdy Z, Jando G, Szabó I, Sik A, Buzsáki G (1995) Sharp wave associated high-frequency oscillation (200 Hz) in the intact hippocampus: network and intracellular mechanisms. J Neurosci 15:30–46
85. Zemankovics R, Veres JM, Oren I, Hájos N (2013) Feedforward inhibition underlies the propagation of cholinergically induced gamma oscillations from hippocampal CA3 to CA1. J Neurosci 33:12337–12351

Chapter 2
Methods

2.1 Experimental Procedures

Chronic recordings were performed in the Langone Medical Center of the New York University and the Department of Physiology of the School of Medicine of the University of Szeged. All experiments were performed in accordance with European Union guidelines (2003/65/CE) and the National Institutes of Health Guidelines for the Care and Use of Animals for Experimental Procedures. The experimental protocols were approved by the Animal Care and Use Committee of New York University Medical Center and the Ethical Committee for Animal Research at the Albert Szent-György Medical and Pharmaceutical Center of the University of Szeged respectively. Animals were anesthetized with isoflurane anesthesia and one or several craniotomies were performed with stereotaxical guidance. One or more silicon probes were mounted in custom-made micro-drives to allow their precise vertical movement after implantation. The probes were inserted over the target region and the micro-drives attached to the skull with dental cement. The craniotomies were sealed with sterile wax. Two stainless steel screws were drilled over the cerebellum and serve as ground and reference for the recordings. Several additional screws were drilled into the skull and covered with dental cement to strengthen the implant. Finally, a copper mesh was attached to the skull with dental cement and connected to the ground screw to act as a Faraday cage and prevent the recording from the environmental electric noise (Fig. 2.1a. For more details, see Vandecasteele et al. [45]). After recovery, the probe is moved gradually in 70–150 μm steps until the desired target is reached. The operated animals were housed in individual cages.

To record neuronal activity during sleep or waking behaviors the probes were connected to a pre-amplifier headstage attached to a long cable pending from the room ceiling that allow full movement to the animal (Fig. 2.1b). The rats' positions during behavioral sessions were estimated using video tracking of two LEDs fixed

A. Fernández-Ruiz, *Extracellular Potentials in the Hippocampus*,
Springer Theses, DOI 10.1007/978-3-319-41039-5_2

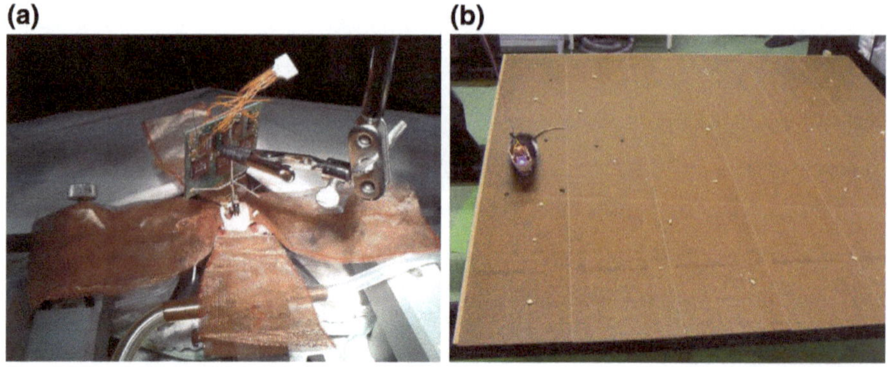

Fig. 2.1 **a** Implantation of a 256 channels silicon probe (NeuroNeuxus) in the hippocampus of a Long Evans rats. Observe the multiplexed pre-amplifier PCB and the microdrive where the probe is mounted. **b** An animal being recorded during exploration for food reward (cookies) in an open field. Two LEDs mounted in the headstage are used for position tracking

to the headstage. The wide-band signal was low-pass filtered and down sampled to 1250 Hz to generate the LFP and was high-pass filtered (>0.8; 20 kHz) for spike detection.

Following the termination of the experiments, the animals were deeply anesthetized, and transcardially perfused first with 0.9 % saline solution followed by 4 % formaldehyde solution. The brains were sectioned by a Vibratome (Leica) at 70 μm sections, parallel with the plane of the implanted silicon probes.

For some of the analysis presented here another dataset was also included. It comprised recordings form the medial entorhinal cortex and hippocampus performed with high-density 32 or 64 electrode (Buzsáki-probes, Neuronexus) while the rats performed different maze tasks or sleep in their home cages. This dataset is publically available at http://crcns.org (hc-3 dataset, [32]).

The experimental data analyzed for the present thesis were not entirely recorded by me but by other researchers from the Buzsáki laboratory (NYU): AntalBerényi, Sean Montgomery, Kenji Mizuseki and John Long.

2.2 LFP Source Decomposition

All the LFP pre-processing and analysis were conducted with custom-made functions or publically available toolboxes in MATLAB (The MathWorks, Inc.).

To address the inverse problem of LFP, that is, to separate the different sources that contribute to the mixed signal, we employed a combination of independent component analysis (ICA) and current source density (CSD) analysis.

2.2.1 Independent Component Analysis

ICA is a blind source-separation technique [4, 16, 26] that can isolate spatially segregated stable patterns of activity in a mixed signal recorded with an array of sensors. Applied to linear profiles of LFPs it can separate physiologically meaningful sources that can be attributed to known anatomical pathways. ICA is able to find the original sources that are statistically independent in a linear mixture. This can be achieved by different ways, thus many different ICA algorithms have been proposed that, although numerically different, are equivalent from a theoretical point of view [15, 25]. We employed the logistic infomax ICA algorithm [4] as implemented in the EEGLAB toolbox (RUNICA; [18] and the fast kernel density ICA algorithm provided by Chen (KDICA, http://cm.bell-labs.com/who//aychen/ica-code.html; [13]). Both algorithms give very similar results.

The application of ICA to a blind source separation problem can be formulated as follows:

Given a linear mixture of n sources $x_1, \ldots, x_n$ that are independent from each other (that is, observing the dynamics of one of them does not give any information about the dynamics of the others), the problem is to separate the source signals given only the mixture

$$x_j(t) = a_{j1}s_1(t) + a_{j2}s_2(t) + \cdots + a_{jn}s_n(t) \tag{2.1}$$

For the following explanation we will eliminate the temporal dimension and denote by x the column vector whose elements are the mixture signals $x_1, \ldots, x_n$ and by$s$ the column vector whose elements $s_1, \ldots, s_n$ are the independent sources:

$$x = A\,s \tag{2.2}$$

A is the mixing matrix with elements a_{ij}. The above equation can also be formulated as

$$x_j = \sum_{i=1}^{n} a_{ji}s_i \tag{2.3}$$

The above equation is called the statistical ICA model. It is a generative model because it describes how the observed mixed signal is generated by an iterative process of mixing of the independent components s_i. The components cannot be directly observed, they are latent variables. Also the mixing matrix $\boldsymbol{A}$ is a priori unknown. We only observe the mixed signal $\boldsymbol{x}$ and have to estimate both $\boldsymbol{A}$ and $\boldsymbol{s}$. After that we can, just by means of an inverse transform, obtain the independent components:

$$\mathrm{s} = \mathrm{Wx} \tag{2.4}$$

where $\boldsymbol{W}$ is the inverse matrix of $\boldsymbol{A}$.

A common first step in the ICA is to sphere or whiten the mixed signals; that is, to remove any existing correlations. If $\boldsymbol{C} = \boldsymbol{E}\{\boldsymbol{xx}'\}$ is the correlation matrix of the mixed signals, the sphering can be accomplished by the linear transformation $\boldsymbol{V} = \boldsymbol{C}^{-1/2}$

$$E\{yy'\} = E\{Vxx'V'\} = C^{\frac{1}{2}} \cdot C \cdot C^{-\frac{1}{2}} = I \tag{2.5}$$

After sphering the independent components can be estimated by and orthogonal transformation of the uncorrelated signals $\boldsymbol{y}$. This is achieved by a rotation of the joint density $p(\boldsymbol{y})$. The appropriated rotation is sought by maximizing the non-Gaussianity of the marginal densities of $p(\boldsymbol{y})$. This relays on the assumption made by ICA (as stated by the central limit theorem) that any linear mixture of independent random variables has necessarily a more Gaussian distribution that the original variables. Most ICA algorithms perform this rotation of the joint density of the signals in an iterative way until it converges. That is achieved when the joint density becomesa product of the marginal densities:

$$p(s) = p(s_1) \cdot p(s_2) \cdots p(s_n) \tag{2.6}$$

The ICA model makes several assumptions that must be fulfilled for the data in order to reach an accurate result [7, 26]: (1) Sources must be stationary. In the context of LFPs, the sources are mainly synaptic transmembrane currents in fixed locations as determined by anatomy [9]. (2) The sources should not have Gaussian-distributed activation strengths, which is the case for brain dynamics [8]. (3) The mixture of the sources must be linear and instantaneous, which can be assumed for electric fields elicited by ionic currents in the extracellular space [34, 36]. (4) The number of sources must be equal or less than the number of sensors, which makes high-density electrode recordings particularly suitable for application of ICA. (5) It is assumed that the observed variable (in this case the LFP) is a mixture of scalar, one dimensional sources. It implies that if the same oscillatory source is observed with a phase delay by different electrodes, then ICA will decompose it into two ICs with a 90 degree phase delay. The temporal activation of the original source would correspond to a linear combination of the time series of those two ICs. Thus, it is important to stress that, though spatially distinct sources which are *perfectly* coherent cannot be properly separated, ICA does not find independent components (ICs) with true temporal independence, and temporal correlations and coherence measures may still be applied to analyze the temporal relationships between the resulting ICs [4, 15, 23, 26, 39].

Our case is that of multiple simultaneous samplings of the mixed signal (the raw LFP recorded by every electrode, Fig. 2.2a). The time series of the LFP recorded by each electrode are the rows of the data matrix $\boldsymbol{D}$. ICA finds the square matrix $\boldsymbol{W}$ (with dimensions equal to the rows of $\boldsymbol{D}$) such that $\boldsymbol{WD} = \boldsymbol{C}$. $\boldsymbol{W}$ is the unmixing matrix because it separates the mixture of signals that is $\boldsymbol{D}$ into its independent sources. $\boldsymbol{C}$ has the same dimensionality as $\boldsymbol{D}$, with each of its rows being the time series of an independent component (Fig. 2.2c, lower panel). Each independent

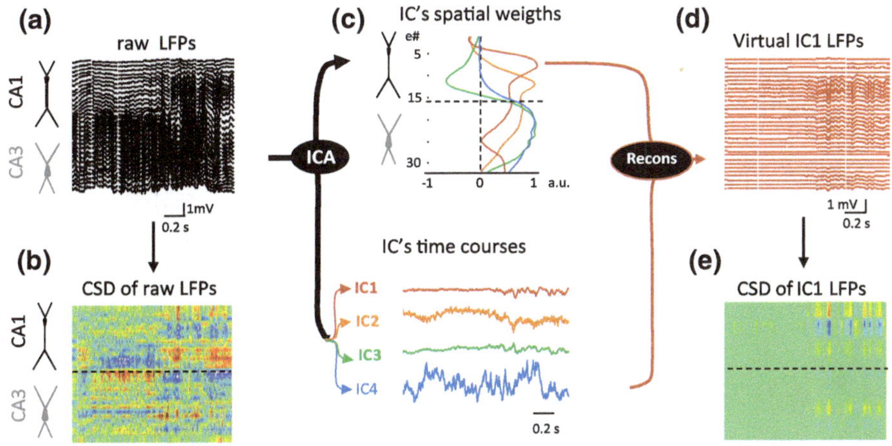

Fig. 2.2 **a** Raw LFPs along the CA1 and CA3 hippocampal subfields (*black* and *gray traces*, respectively). **b** CSD of the LFPs renders a complex mixture of currents as expected for multiple synaptic inputs. **c** ICA of LFPs provides four main ICs, each defined by the curve of spatial weights (*top panel*) and a time course (*bottom traces*). **d** Reconstructed (virtual) LFPs for IC4. **e** CSD of the virtual LFPs provides precise spatiotemporal maps of inward/outward currents for unique spatially coherent synaptic input. Modified from [20]

component is obtained by multiplying each sampled signal by each row of $\boldsymbol{W}$ (the so called unmixing functions or ICA filters). This process can be also view as solving the inverse problem of the LFP; that is estimate the sources giving only the potentials distribution.

The inverse of the mixing matrix $\boldsymbol{W}$ that transforms the LFP data into the ICs gives the channel weight of each component that is captured for each sensor. When plotted according to the anatomical location of the electrodes, this corresponds to the spatial voltage loadings of each IC (Fig. 2.2c, upper panel). We ranked the components by the amount of variance they explain in the original data (*relative power*). Once ICs have been extracted from the raw LFP traces, they can be analyzed as if they were active independently from activities at other locations. We reconstructed the virtual LFP produced by a single IC by multiplying the IC time course by its correspondent voltage loading (Fig. 2.2d). For each component a, that is achieved by

$$W_a^{-1} C_a = D_a \tag{2.7}$$

That is also known as solving the forward problem, or reconstructing the potentials distribution given the current sources.

Before application of the ICA algorithm, we performed several pre-processing steps. For hippocampal LFPs, there are only a small number of physiologically meaningful ICs with significant amplitude and identifiable spatial loadings [5, 21–23, 39]. Before applying ICA, we therefore employed a principal component analysis

(PCA) reduction maintaining 98.0 % of the original LFP variance. This process aids in the convergence of ICA to stable components, and results in a smaller number of ICs. We also whitened the data before applying ICA to reduce the computational complexity of the analysis while maintaining its statistical consistency [12, 25].

Due to the parallel anatomical arrangement of the principal cells and the stratification of afferent axon terminals, the hippocampus is especially well suited for ICA decompositions of its LFPs. However, only those synaptic inputs with enough postsynaptic current, synchrony, and spatial clustering are suitable for ICA separation; thus, very weak or sparse currents are not easily discernible.

2.2.2 Current Source Density Analysis

The traditional approach to solve the inverse problem of the LFPs is performing current source density (CSD) analysis. CSD analysis [24, 31] determines the magnitude and location of the net transmembrane currents generated by neuronal elements within a small volume of tissue.

In a macroscopic level the transmembrane current density I_m per unit length is related with the extracellular current density $\boldsymbol{J}$ through the divergence:

$$\nabla \boldsymbol{J} = I_m \tag{2.8}$$

If we assume a quasistatic description of the electric field in the extracellular medium (neglecting capacitive and inductive effects), Ohm law can be applied,

$$\boldsymbol{J} = \sigma \boldsymbol{E} \tag{2.9}$$

where σ is in general the conductivity tensor.

The relation between the electric field $\boldsymbol{E}$ and the extracellular potential Φ

$$\boldsymbol{E} = -\nabla \Phi \tag{2.10}$$

allows us to establish a linear dependency between the extracellular current density $\boldsymbol{J}$ and the gradient of the electric potential in the medium

$$\boldsymbol{J} = -\sigma \nabla \Phi \tag{2.11}$$

Substituting this expression of $\boldsymbol{J}$ into the current conservation equation, we get the Poisson equation, which establishes the relation between the electric potential and the volumetric current density

$$\nabla(\sigma \nabla \Phi) = -I_m \tag{2.12}$$

In the simplest approach, we employed a 1D approximation by calculating the second spatial derivative of the LFP profiles (Fig. 2.2b), and calculated the CSD according to a central differences formula,

$$CSD(z_i) = -\sigma \frac{\Phi(z_i + \delta) - 2\Phi(z_i) + \Phi(z_i - \delta)}{\delta^2} \tag{2.13}$$

where δ is the distance between electrodes.

This approach assumes isotropy and negligible net contributions in the XY-plane perpendicular to the cellular axis. That is suitable for laminated structures with parallel arrangement of principal cells, as is the case of CA1, if the recording electrodes are placed parallel to the main cell axis (z) and an homogeneous population of cells is synchronously active (as for example during evoked potentials). In this traditional approach to CSD estimation it is also assumed homogeneous resistivity.

However, the above assumptions are not fulfilled in most real cases. When the cells in a small volume of tissue are active only in a given time there can be a significant current spreading in the x and y directions. This may result in an underestimation of true sink and sources as well as in the occurrence of spurious ones. To address those issues we employed the recently developed inverse CSD (iCSD) method [35]. This method consists of first define a forward model to describe the potentials that are produced by localized current sources and then invert this model by means of a numerical matrix inversion to allow direct calculation of localized discrete sources form the measured potentials distribution. The solution of the forward model is given by the following equation

$$\Phi(r_e) = \frac{1}{4\pi\sigma} \iiint_V \frac{CSD(r)}{|r - r_e|} d^3 r \tag{2.14}$$

In most LFP studies the application of CSD analysis does not take into account tissue inhomogeneity and anisotropy effects, assuming a negligible contribution of these effects [9, 19]. In the present work we also adopted this approach; however we took advantage of the iCSD method toexplicitly incorporate a priori knowledge about the geometry of the sources, such as the volume of activated tissue and the relative position of the sources to the recording electrodes.

In addition to the raw LFP, we performed CSD analysis of the reconstructed virtual LFP produced by a single IC, which renders a map of the current distribution of a single anatomical input or LFP source [20, 21, 39].

2.3 Time-Frequency Analysis of LFPs

To characterize the LFP signals in the spectral domain we employ a multi-taper implementation of the fast Fourier transform ([30]; Chronux toolbox, http://www.chronux.org). Continuous data were segmented in 1–5 s epochs for all spectral analysis. Spectral power was estimated with a Hanning window and averaged across all data epochs.

In the multi-taper power spectrum method employed here a set of independent estimates of the power spectrum are computed, by multiplying the signal by orthogonal tapers (windows) which are constructed to minimize the spectral leakage due to the finite length of the data set. The tapers are the discrete set of eigenfunctions that solve the variational problem of minimizing leakage outside of a predefined frequency band. Once the tapers $\boldsymbol{w_k}(t)$ are computed for a chosen frequency bandwidth, the total power spectrum $\boldsymbol{P_X}$ can be estimated by averaging the individual spectra given by each tapered version of the time series $\boldsymbol{x}(t)$; the kth eigenspectrum $\boldsymbol{X_k}$ is the discrete Fourier Transform of $\boldsymbol{x}(t)\boldsymbol{w_k}(t)$

$$P_x(f) = \frac{\sum_{k=1}^{K} \mu_k |X_k(f)|^2}{\sum_{k=1}^{K} \mu_k} \tag{2.15}$$

where μ_k are the eigenvalues of w_k.

This procedure yields a better and more stable estimate of a signal power spectrum than single taper methods.

Two basic measures of pairwise synchronization were employed: cross-correlation and coherence. The cross-correlation function is a measure of the linear covariance between two signals $\boldsymbol{x}$and $\boldsymbol{y}$and can be estimated as follows,

$$cc_{xy} = \frac{\sum_{k=1}^{n} (x_k - \bar{x})(y_k - \bar{y})}{\sqrt{\sum_{k=1}^{n} (x_k - \bar{x})^2} \cdot \sqrt{\sum_{k=1}^{n} (y_k - \bar{y})^2}} \tag{2.16}$$

where $\bar{x}$ and $\bar{y}$ are the means of the respective signals.

This method has the advantages of being a straightforward method that yields a normalized value giving a gross indication of the degree of similarity of two signals (i.e. LFPs recorded at separate locations). However, for a frequency resolved estimate of the linear covariance of two signals we need to employ the coherence.

$$coh_{xy}(f) = \sqrt{\frac{|\langle C_{xy}(f) \rangle|^2}{|\langle C_{xx}(f) \rangle| \cdot |\langle C_{yy}(f) \rangle|}} \tag{2.17}$$

where $\boldsymbol{C_{xx}}(f)$ is the cross-spectral density for the frequency f between $\boldsymbol{x}$ and $\boldsymbol{y}$, and $\langle \ldots \rangle$ indicates averaging over segments.

To assess spectral events at a high resolution in time and frequency, the complex wavelet transform (CWT) of the LFP was calculated using complex Morlet wavelets [42]. The CWT gives amplitude and phase measures for each wavelet scale at all time points in the data, obtained by convolving the real and imaginary parts (which are phase shifted by 90°) of the wavelets with the data vectors.

A wavelet is a scalable function with zero mean, well localized in time.

$$\int_{-\infty}^{\infty} \psi(t)dt = 0 \tag{2.18}$$

A family of wavelets can be constructed from a "mother" function $\Psi(t)$, which is confined to a finite interval, translated with a factor u and expanded with a scale parameter s,

$$\psi_{u,s}(t) = \frac{1}{\sqrt{s}} \cdot \psi\left(\frac{t-u}{s}\right) \tag{2.19}$$

Then the wavelet analysis of a signal x can be performed by

$$\langle x(t), \psi_{u,s}(t) \rangle = \int x(t)\psi_{u,s}(t)dt \tag{2.20}$$

In our case, we employed the Morlet complex waveform,

$$\psi(t) = \left(e^{i\omega_0 t} - e^{-\omega_0^2 t^2/2}\right) \cdot e^{-t^2/2\sigma_t^2} \tag{2.21}$$

where ω_0 is the center frequency of the wavelet and σ_t a bandwidth parameter determining its rate of decay. The width or number of cycles of the wavelet is given by $\omega_0\sigma_t^2$.

If this complex wavelet is convolved with the signal x we get the CWT of that signal,

$$W(t) = (\psi \circ x)(t) = \int \psi(t') \times (t - t')\, dt' = \mathrm{A}^W(t)e^{i\phi_W(t)} \tag{2.22}$$

where $\phi_W(t)$ is the phase for each time sample. The Morlet mother waveform has Gaussian modulation in both time and frequency, thus offering optimal resolution in both domains.

The simpler Hilbert transform was also employed in some cases (as when only the theta oscillations were considered) to extract the phase or amplitude of LFP signals. The phase $\phi_W(t)$ and amplitude $A_x(t)$ are given respectively by the argument and modulus of the complex analytical signal:

$$\xi(t) = x(t) + ix_H(t) = A_x(\mathrm{t})\mathrm{e}^{i\varphi_x(t)} \tag{2.23}$$

where $x_H(t)$ is the Hilbert transform of $x(t)$, defined as:

$$x_H(t) = \frac{1}{\pi} P.V. \int_{-\infty}^{\infty} \frac{x(t')}{t - t'} dt' \tag{2.24}$$

with $P.V.$ denoting the Cauchy principal value.

Both wavelet and Hilbert transforms give very similar results in all cases.

The phase-amplitude cross-frequency coupling (CFC) during theta oscillations for a given LFP recording was assessed using the modulation index (MI) introduced by Tort et al. [44]. We took the phase of the LFP recorded at the CA1 pyramidal cell layer and the amplitude of either LFP recorded in different layers or the time course of different LFP-generators, in all cases the procedure was exactly the same. The raw LFP signal (Fig. 2.3a) was band-pass filtered in the low frequency band (Fig. 2.3b) and the phase of the analytic signal given by the Hilbert transform or CWT was calculated (Fig. 2.3c). Amplitude of the filtered signal in the broad gamma band (30–300 Hz, Fig. 2.3d) was also obtained from the CWT. The MI was calculated by measuring the divergence of the observed amplitude distribution from the uniform distribution (Fig. 2.3e). Comodulogram phase-amplitude plots were constructed representing in pseudocolor scale the MI values of multiple phase-amplitude frequency pairs (Fig. 2.3f). The statistical significance of the MI values (P-value) was assessed by a surrogate analysis (n = 1000 surrogates) with random shifts between the phase and amplitude time series [11].

2.4 Single Unit Analysis

Neuronal spikes were detected from the digitally high-pass filtered LFP (0.8–5 kHz) by a threshold crossing-based algorithm (*Spikedetekt2*; https://github.com/klusta-team/spikedetekt2). Detected spikes were automatically sorted using the masked EM algorithm for Gaussians mixtures implemented in KlustaKwik ([27]; https://github.com/klusta-team/klustakwik/), followed by manual adjustment of the clusters using the KlustaViewa software ([37]; https://github.com/klusta-team/klustaviewa/; Fig. 2.4) to get well-isolated single units. Multiunit or noise clusters where discarded for the analysis. Putative pyramidal cells and interneurons were separated on the basis of their autocorrelograms and waveforms characteristics [17, 33, 40], assisted by monosynaptic latency excitatory and inhibitory interactions between simultaneously recorded, well-isolated units [2, 33]. Kenji Mizuseki at the Buzsáki laboratory performed most of the unit clustering and classification.

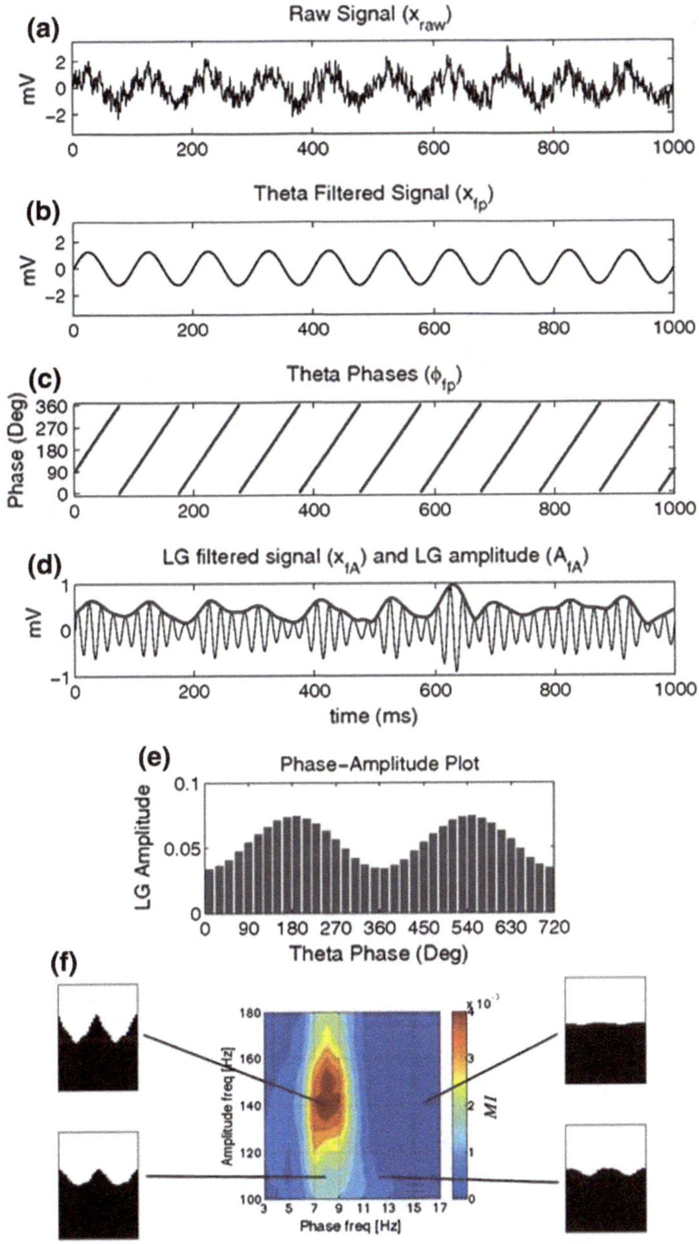

Fig. 2.3 **a** Example of raw LFP signal. After filtering in the theta band (**b**), the phase is calculated using the Hilbert transform. **c** The same or other signal is filtered in the frequency band of interest and its amplitude (**d**) is obtained from the CWT. The mean amplitude distribution over theta phase is then calculated (**e**). The phase-amplitude MI is obtained by measuring the divergence of the observed amplitude distribution from the uniform distribution. A phase-amplitude comodulogram plot is constructed representing in pseudocolor scale the obtained MI values for multiple phase-amplitude pairs (in this case the phase was not only calculated for the theta filtered signal but for a range of frequencies). Modified from [43, 44]

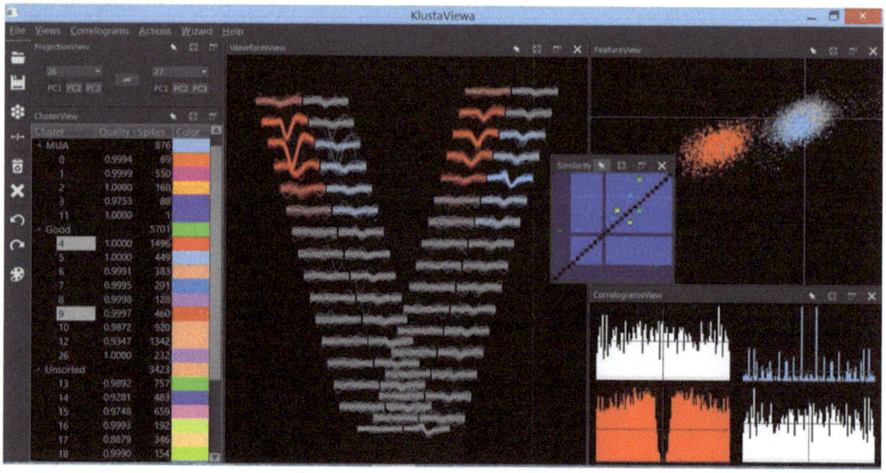

Fig. 2.4 Screen capture from KlustaViewa, an open-source software for manual clustering of neuronal spikes. In the *central panel*, two single units with different anatomical location are displayed. Observe the different features provided for assisting the process: auto and cross-correlograms. PCA projections, similarity matrix. From [37]

The phase-locking of spikes to the LFP was measured for individual units using the wavelet or Hilbert phase at the time of each spike. Modulation indices were calculated using the mean resultant length of the phases, and significance was estimated applying the Rayleigh test for non-uniformity using the circular statistics toolbox provided by Berens [6]. Erik Schomburg at the Buzsáki laboratory implemented and performed unit-LFP analyses.

2.5 Modeling of LFPs with Finite Elements Method

Multiple physical problems can be mathematically formulated as a partial derivatives differential equation. As a general rule, those differential equations are very hard to solve. Only in those cases in which it is possible to make simplifications on the dimensionality and geometry of the problem an analytical solution is available. However, in most of the cases analytical methods are not feasible or are inexact (as can be the case of a tridimensional electromagnetic problem in a complex geometry). For solving this kind of problems numerical models are necessary and useful. Those methods discretize the differential equation into a linear system of equations, solving it in an iterative way. One of those methods is the analysis by means of finite elements (FEM).

The basis of FEM is to divide the geometry in which a differential equation of a scalar or vector field needs to be solved (i.e. the electric potential) in small elements, typically tetrahedral. In each element field equations determined by a variationalprinciple and local sources are solved. Boundary conditions are applied and thus it is possible to obtain the algebraic system from which a solution of the problem is found. FEM allows explicit models of the geometry and electrical properties of the brain to be generated and to solve potential and current distributions in such a complex structure over time. In our case, the physical problem that needs to be solved with FEM can be described as follows. The initial condition is a certain distribution of volumetric current sources in a tridimensional complex structure (i.e. the rat hippocampus). Each of the sources is simultaneously activated with different time series. As a result of the uneven spatiotemporal varying summation of currents in the volume, an electric potential distribution arise. Thus, our goal is to calculate this potential distribution for each spatial point at anytime.

As such, we modeled the rat hippocampus establishing the geometry of current sources and their temporal activation, and fixed the boundary conditions. Two different models were built, one of the whole dorsal hippocampus in its planar section, and another of just the dentate gyrus region. The structure was enclosed in a larger volume simulating the extracellular space. This outer compartment was sufficiently large so asnot to distort the field lines in the central region of interest. We tested different surrounding volumes (up to four times longer in each direction) while maintaining boundary conditions. The chosen volume rendered LFPs that were at least 90 % of the maximum amplitude obtained with the largest volume (>95 % in most tested points). For the sake of simplicity, the tissue was considered to be purely resistive, isotropic and homogeneous [29] but see [3]. Conductivity was set as 0.3 S/m and relative permittivity as 10^6. FEM current sources were defined as tridimensional volumes, allowing the electrical currents produced by multiple synchronously activated neurons to be compiled in a few block-like current generators that jointly obey the principle of charge conservation. This is an important advantage, which makes it suitable to reproduce LFPs in a volume generated by activated sections of layered structures made up of neurons arranged in parallel as is the case of the hippocampus or cortex. Accordingly, the size and geometry of the blocks of current represent the physical extension of the synchronously activated dendritic domains, i.e., the portion of the population of target cells that elicits postsynaptic transmembrane currents upon coherent activation of a group of afferent axons (Fig. 2.5a).

We found that eighth of these blocks with 100 μm thickness in the case of CA1 and four 60 μm thickness block for each blade of the DG were sufficient to reproduce the main laminar features of hippocampal LFPs [23]. In the case of CA1, one for the basal dendrites, one for the pyramidal cell body layer and six for the apical dendrites. For the DG, one block represented the granular cell body layer and three the dendrites. These blocks were bent to replicate the curved geometry of the hippocampus and divided into approximately 200 μm sections that could be independently activated to analyze the effects of the spatial coherence and synchrony of the inputs. For simplicity, we used non-overlapping blocks of current, representing inputs with imaginary topological projection of varying synaptic territories.

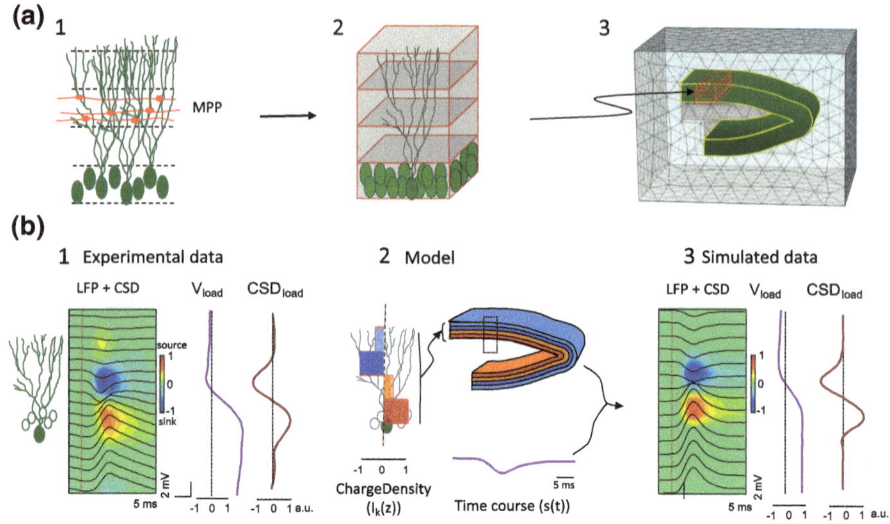

Fig. 2.5 **a** The gross cytoarchitecture of the granule cell (GC) population (*1*) was assembled as four stacked rectangular blocks each representing a subcellular domain. *2* The blocks were given appropriate curvature and dimensions to reproduce a stereotyped U-shaped geometry of the dorsal dentate gyrus (DG). *3* A tetrahedron adaptive recording mesh simulating the conductive extracellular medium was built in and around the cellular component and was large enough not to distort electrical fields. **b** The block sources were activated using real time activations. In this example, we used the excitatory medial perforant pathway (MPP) that makes synaptic contact in the middle third of the dendritic tree (see **a***1*). Total charge was balanced across all blocks at every instant. Charge was distributed throughout all four compartments according to weights obtained in the spatial map of CSD analysis for the electrical activation of this pathway. V_{load} and CSD_{load} represent the spatial weights of voltage and CSD along the GC main axis. The excitatory sink (in blue) is surrounded by a strong passive source (warm colors) in cell soma and a weaker source in distal dendrites. *2* At all times, the blocks received proportional charge density with predefined polarity and identical time course. *3* Example of FEM simulated data. Adapted from [23]

In the present models we deliverately excluded any possible contribution of the extracellular currents from other cellular types apart from pyramidal and granular cells (e.g. interneurons or glia). Most types of interneuros and astrocites have multipolar dendritic trees [1] and hence, a 'closed-field' extracellular configuration of the electric field is established by their synaptic activation (Lorente de Nó, 1947). Consequently, their currents do not (or they only poorly) spread beyond their physical limits [10, 19, 28]. In consequence, these cells do not significantly add their currents in the extracellular space and they only negligibly contribute to the LFP (even if they are strongly synchronized).

We applied Dirichlet boundary conditions by setting the field to the ground value on the external surface of the enclosing volume and imposing charge conservation inside the total volume. A tetrahedral adaptive grid of the highest resolution (smallest size, 0.05 μm) was used to ensure the correct resolution of field equations in the curved compartments (Fig. 2.5a3).

The FEM approach is widely used in magnetoencephalography and scalp electroencephalography [14, 38, 41], although its application to the study of LFPs is almost inexistent. We used a commercial FEM software tool, COMSOL Multiphysics® (www.comsol.com).

Time-dependent analyses were performed using the AC/DC module of COMSOL for all nodes in the extracellular and cellular volume. Thespatial and temporal dynamics of LFP and CSD distributions were evaluated after activation of individual subcellular domains by different inputs (the rationale of the performed simulations is illustrated in Fig. 2.5b; see [23]. The temporal activation of a particular synaptic afference, such as the MPP excitatory input, was taken as the input signal, $s(t)$. For simplicity, we represent here an MPP-evoked fEPSP. The axons of this pathway establish synaptic contact with GCs in the middle third of their dendritic tree (Fig. 2.5b1, green axons). The spatial distribution of the CSD corresponding to this activation (Fig. 2.5b1, contour map) was compartmentalized into four spatial blocks that jointly configured the GC population, such that they roughly reproduced the same spatial profile (Fig. 2.5b2). The sum of charge densities was set to zero, as imposed by current conservation law. The electric fields and potentials elicited by these currents were calculated for the entire tissue volume by FEM.

Linear profiles of simulated LFPs comparable to those recorded in vivo were built using several linear tracks along the vertical z-axis, which contained up to 40 registration points spaced at 50 μm intervals and that were placed in the middle of the structure to produce the most homogenous field contribution. LFP profiles were constructed from the instantaneous voltage signal recorded at each simulated recording point.

References

1. Amaral DG (1978) A Golgi study of cell types in the hilar region of the hippocampus in the rat. J Comp Neurol 182:851–914
2. Bartho P, Hirase H, Monconduit L, Zugaro M, Harris KD, Buzsàki G (2004) Characterization of neocortical principal cells and Interneurons by network interactions and extracellular features. J Neurophysiol 92:600–608
3. Bédard C, Destexhe A (2011) Generalized theory for current-source-density analysis in brain tissue. Phys Rev E 84:041909
4. Bell AJ, Sejnowski TJ (1995) An information-maximization approach to blind separation and blind deconvolution. Neural Comput 7:1129–1159
5. Benito N, Fernandez-Ruiz A, Makarov VA, Makarova J, Korovaichuk A, Herreras O (2014) Spatial modules of coherent activity in pathway-specific LFPs in the hippocampus reflect topology and different modes of presynaptic synchronization. Cereb Cortex 24:1738–1752
6. Berens P (2009) CircStat: a MATLAB toolbox for circular statistics. J Stat Softw 31
7. Brown GD, Yamada S, Sejnowski TJ (2001) Independent component analysis at the neural cocktail party. Trends Neurosci 24:54–56
8. Buzsàki G, Mizuseki K (2014) The log-dynamic brain: how skewed distributions affect network operations. Nat Rev Neurosci 15:264–278

9. Buzsáki G, Wang X-J (2012) Mechanisms of gamma oscillations. Annu Rev Neurosci 35:203–225
10. Buzsáki G, Anastassiou CA, Koch C (2012) The origin of extracellular fields and currents—EEG, ECoG, LFP and spikes. Nat Rev Neurosci 13:407–420
11. Canolty RT, Edwards E, Dalal SS, Soltani M, Nagarajan SS, Kirsch HE, Berger MS, Barbaro NM, Knight RT (2006) High gamma power is phase-locked to theta oscillations in human neocortex. Science 313:1626–1628
12. Chen A, Bickel PJ (2005) Consistent independent component analysis and prewhitening. IEEE Trans Sig Proc 53:3625–3632
13. Chen A (2006) Fast kernel density independent component analysis. Lect Notes Comput Sci 3889:24–31
14. Chen M, Mogul DJ (2009) A structurally detailed finite element human head model for simulation of transcranial magnetic stimulation. J Neurosci Meth 179:11–120
15. Choi S, Cichocki A, Park HM, Lee SY (2005) Blind source separation and independent component analysis: a review. Neur Inf Process Lett Rev 6:1–57
16. Comon P (1994) Independent component analysis, a new concept? Sig Process 36:287–314
17. Csicsvari J, Hirase H, Czurko A, Buzsáki G (1998) Reliability and state dependence of pyramidal cell-interneuron synapses in the hippocampus: an ensemble approach in the behaving rat. Neuron 21:179–189
18. Delorme A, Makeig S (2004) EEGLAB: an open source toolbox for analysis of single-trial EEG dynamics including independent component analysis. J Neurosci Methods 134:9–21
19. Einevoll GT, Kayser C, Logothetis NK, Panzeri S (2013) Modelling and analysis of local field potentials for studying the function of cortical circuits. Nat Rev Neurosci 14:770–785
20. Fernández-Ruiz A, Herreras O (2013) Identifying the synaptic origin of ongoing neuronal oscillations through spatial discrimination of electric fields. Front Comput Neurosci 7:5
21. Fernández-Ruiz A, Makarov VA, Benito N, Herreras O (2012) Schaffer-specific local field potentials reflect discrete excitatory events at gamma frequency that may fire postsynaptic hippocampal CA1 units. J Neurosci 32:5165–5176
22. Fernández-Ruiz A, Makarov VA, Herreras O (2012) Sustained increase of spontaneous input and spike transfer in the CA3-CA1 pathway following long-term potentiation in vivo. Front Neural Circuits 6:71
23. Fernández-Ruiz A, Muñoz S, Sancho M, Makarova J, Makarov VA, Herreras O (2013) Cytoarchitectonic and dynamic origins of giant positive local field potentials in the dentate gyrus. J Neurosci 33:15518–15532
24. Freeman JA, Nicholson C (1975) Theory of current source-density analysis and determination of conductivity tensor for anuran cerebellum. J Neurophysiol 38:356–368
25. Hyvärinen A, Oja E (2000) Independent component analysis: algorithms and applications. Neural Netw 13:411–430
26. Hyvärinen A, Karhunen J, Oja E (2004) Independent component analysis. Wiley, New York
27. Kadir SN, Goodman DFM, Harris KD (2014) High-dimensional cluster analysis with the masked EM algorithm. Neural Comput 26:11
28. Lindén H, Tetzlaff T, Potjans TC, Pettersen KH, Gru S, Diesmann M, Einevoll G (2011) Modeling the spatial reach of the LFP. Neuron 5:859–872
29. Logothetis NK, Kayser C, Oeltermann A (2007) In vivo measurement of cortical impedance spectrum in monkeys: implications for signal propagation. Neuron 55:809–823
30. Mitra PP, Pesaran B (1999) Analysis of dynamic brain imaging data. Biophys J 76:691–708
31. Mitzdorf U (1985) Current source-density method and application in cat cerebral cortex: investigation of evoked potentials and EEG phenomena. Physiol Rev 65:37–100
32. Mizuseki K, Diba K, Pastalkova E, Teeters J, Sirota A, Buzsáki G (2014) Neurosharing: large-scale data sets (spike, LFP) recorded from the hippocampal-entorhinal system in behaving rats. F1000 Research 3
33. Mizuseki K, Sirota A, Pastalkova E, Buzsáki G (2009) Theta oscillations provide temporal windows for local circuit computation in the entorhinal-hippocampal loop. Neuron 64: 267–280

34. Nunez PL, Srinivasan R (2006) Electric fields of the brain: the neurophysics of EEG, 2nd edn. Oxford University Press, New York
35. Pettersen KH, Devor A, Ulbert I, Dale AM, Einevoll GT (2006) Current-source density estimation based on inversion of electrostatic forward solution: effects of finite extent of neuronal activity and conductivity discontinuities. J Neurosci Methods 154:116–133
36. Plonsey R, Heppner DB (1967) Considerations of quasi-stationarity in electrophysiological systems. Bull Math Biophys 29:657–664
37. Rossant C, Kadir SN, Goodman DF, Schulman J, Hunter ML, Saleem AB, Grosmark A, Belluscio M, Denfield GH, Ecker AS, Tolias AS, Solomon S, Buzsáki G, Carandini M, Harris KD (2016) Spike sorting for large, dense electrode arrays. Nat Neurosci 19:634-41
38. Salvador R, Silva S, Basser PJ, Miranda PC (2011) Determining which mechanisms lead to activation in the motor cortex: a modeling study of transcranial magnetic stimulation using realistic stimulus waveforms and sulcal geometry. Clin Neurophysiol 122:748–758
39. Schomburg EW, Fernández-Ruiz A, Berényi A, Mizuseki K, Anastassiou CA, Koch C, Buzsáki G (2014) Theta phase segregation of input-specific gamma patterns in entorhinal-hippocampal networks. Neuron 84:470–485
40. Stark E, Roux L, Eichler R, Senzai Y, Royer S, Buzsáki G (2014) Pyramidal cell-interneuron interactions underlie hippocampal ripple oscillations. Neuron 83:467–8
41. Thielscher A, Opitz A, Windhoff M (2011) Impact of the gyral geometry on the electric field induced by transcranial magnetic stimulation. Neuroimage 54:234–243
42. Torrence C, Compo G (1998) A practical guide to wavelet analysis. B Am Meteorol Soc 79:61–78
43. Tort ABL, Komorowski R, Eichenbaum H, Kopell N (2010) Measuring phase-amplitude coupling between neuronal oscillations of different frequencies. J Neurophysiol 2:1195–1210
44. Tort ABL, Kramer MA, Thorn C, Gibson DJ, Kubota Y, Graybiel AM, Kopell NJ (2008) Dynamic cross-frequency couplings of local field potential oscillations in rat striatum and hippocampus during performance of a T-maze task. Proc Natl Acad Sci USA 105:20517–20522
45. Vandecasteele M, M S, Royer S, Belluscio M, Berényi A, Diba K, Fujisawa S, Grosmark A, Mao D, Mizuseki K, Patel J, Stark E, Sullivan D, Watson B, Buzsáki G (2012) Large-scale recording of neurons by movable silicon probes in behaving rodents. J Vis Exp 4:e3568

Chapter 3
Current Sources of Hippocampal LFPs

One of the ultimate goals of the investigations on neural circuit dynamics is to understand the input-output transformation of neuronal signals, i.e., how neuronal activity in an upstream region affects the firing rate and spike timing in neurons of a downstream region. Unfortunately, studying LFP signals recorded from a single or few sites as they are most commonly recorded, cannot properly address the problem of input-output transformation because the LFP signal recorded at any given site represents a weighted sum of multiple neuronal sources in unknown proportions [10, 15]. Ideally, one would like to decompose this macroscopic signal into its individual sources and relate them to the output spiking of neurons to reveal the relative influence of the individual inputs to spike outputs during different behaviors.

Information extracted from LFP signals can be improved significantly by monitoring multiple sites at high spatial resolution [13, 14]. Since afferents to dendrites in the hippocampus are spatially segregated, their behavior-dependent contributions can be separated by sufficiently high density sampling of the LFP [7, 11, 17, 34].

In the first part of the present work, I characterize the main current sources of hippocampal LFPs. For this purpose, LFP recordings were performed with high-density silicon linear probes (either single-shank or 8 shanks probes with 32 electrodes spaced 50 μm; Fig. 3.1a) covering the full transversal axis of the hippocampus of behaving rats. Such probes allow usto record LFPs and unit activity in all hippocampal layers and subregions. The electrodes were slowly advanced during the days following the implantation until reaching the final position spanning from the CA1 stratum oriens to the DG lower blade or CA3 str. oriens. Once finalized the experiment the position of the electrodes was histologically verifiedand compared with the electrophysiological data (Fig. 3.1b).

A. Fernández-Ruiz, *Extracellular Potentials in the Hippocampus*,
Springer Theses, DOI 10.1007/978-3-319-41039-5_3

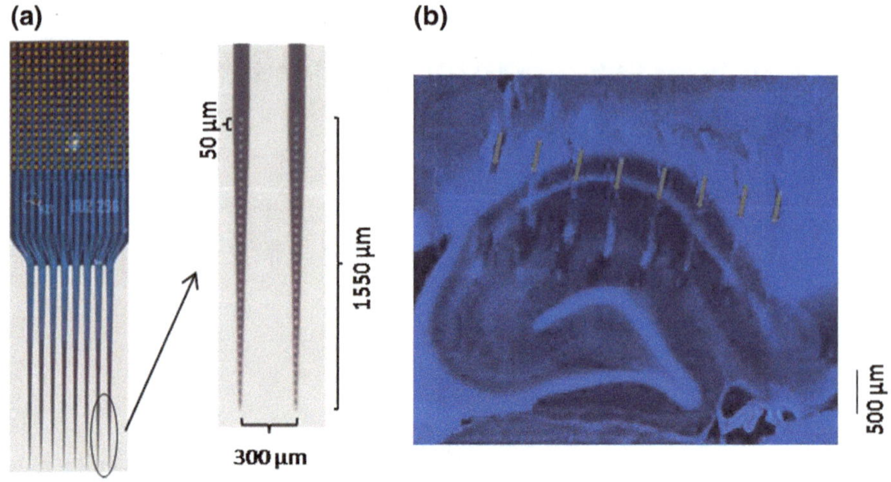

Fig. 3.1 **a** Silicon probe employed for hippocampal recordings (*Neuronexus*). It has 8 *shanks* separated 300 µm each with 32 *electrodes* spaced 50 µm. Thus, it covers 1550 µm in depth and 2100 µm in extension. **b** Final position of the probe was verified histologically after perfusion of the animal. Transversal section of the dorsal *hippocampus* stained with DAPI. Modified from Schomburg et al. [39]

3.1 Experimental Investigation

3.1.1 CA1 Region

Linear recordings along the vertical axis of the hippocampus clearly show the characteristic laminar variations of LFPs (Fig. 3.2a). The hippocampus, and specially the CA1 region, has a great advantage for the study of LFPs due to the parallel arrangement of their cells and the stratification of the synaptic inputs. This anatomical organization results in layer-specific LFP patterns elicited by the inputs innervating the pyramidal or granular cells in restricted dendritic domains [4–7, 18, 34].

ICA discriminates the contributing sources to the LFP based on their distinct spatial distribution. When applied to the multi-electrode wide-band LFP signals of the CA1 region, ICA found three major pathway-specific independent-components (ICs). By convoluting the LFP with the inverse of the mixing matrix estimated by ICA we get the relative voltage weight of every IC in each electrode. Projecting those weights to the anatomical space (i.e., the spatial arrangement of the electrodes in the tissue) we obtain the spatial voltage loading in the dorsoventral axis (z) of each IC (Fig. 3.2b). The second derivative in the z-axis of those voltage loadings represents the CSD loading of this particular IC (Fig. 3.2c), that is the transmembrane currents elicited by this particular input in the target population.

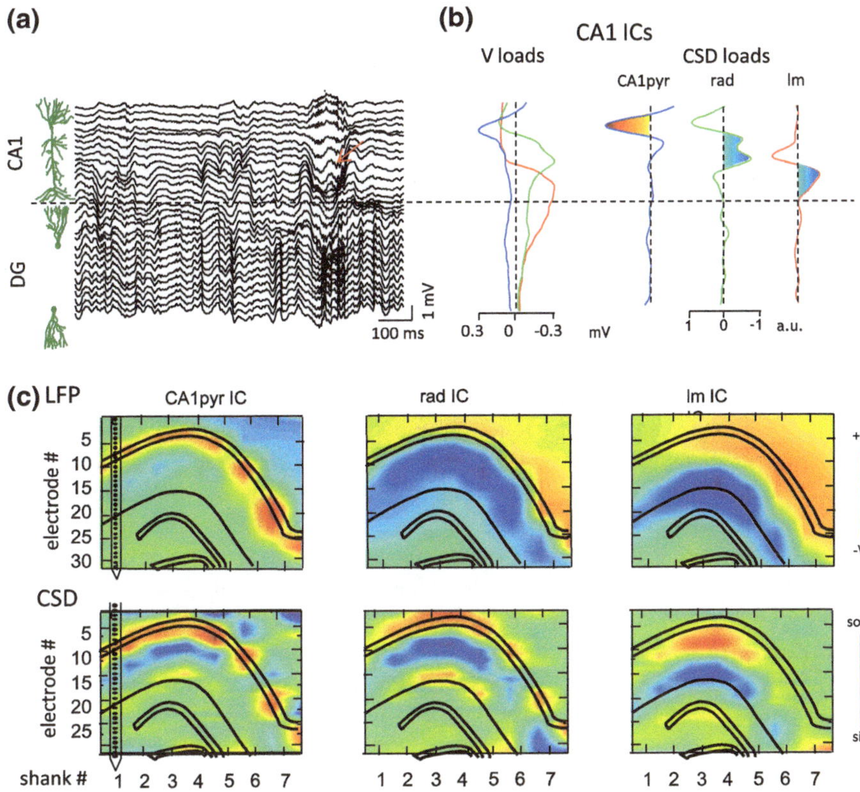

Fig. 3.2 **a** LFP profile along CA1 and DG displaying some characteristic CA1 LFP patterns, sharp wave ripples (*red arrow*). **b** Three main ICs were found for CA1 LFPs with largest voltage and active currents at different layers: the pyramidal layer (*CA1pyr*), str. radiatum (*rad*) and str. lacunosum-moleculare (*lm*). **c** 2D voltage and CSD distributions for the 3 ICs highlight their layer-specific distributions. Modified from Schomburg et al. [39]

The maximal amplitude of the first component (*rad*) is at the level of the str. radiatum and its CSD depth profile matched the source–sink–source distribution of the spontaneously occurring sharp-waves during immobility (Fig. 3.2a, b; green traces; [5, 6, 18, 19, 35]), corresponding to the apical dendritic excitation of CA1 pyramidal neurons (manifested as a strong sink in the str. radiatum) by the synchronous CA3 output [12]. Another IC (*lm*) peaked below the first component, corresponding to the str. lacunosum-moleculare-related current sink (Fig. 3.2a, b red traces; [4, 7]). The peak amplitude of the third component (*CA1pyr*) occurred at the depth of CA1 pyramidal layer (Fig. 3.2a, b; blue traces), also identified by the large amplitude ripples and unit firing [33]. This IC is characterized by a prominent current source centered at the pyramidal layer.

The above results were obtained by applying ICA to single-shank recordings but I also applied it to the 2D matrix of 256 electrode arrays (8 shanks separated 300 μm).

By doing so, the same three main ICs were obtained in the CA1 region. When plotted in two dimensions they display clear layer-specific distributions (Fig. 3.2c). The *CA1pyr* IC shows larger positive voltage along the CA1 pyramidal layer and the CSD map reveal a source surrounded by smaller sinks. This distribution matches the expected from a perisomatic inhibition and its passive return currents in apical and basal dendrites. The *radiatum* component displays larger negative voltage along the CA1 str. radiatum and a polarity reversal in the pyramidal layer. The CSD map reveals the expected strong sink in the dendritic domain of the Schaffer collaterals (CA3 to ipsilateral CA1 input) flanked by sources in the str. lacunosum-moleculare and pyramidal layer. The *lac-mol* IC has larger negative voltage around the hippocampal fissure (the separation between CA1 and DG) but the CSD analysis reveals a current dipole restricted to CA1 distal dendrites, the dendritic domain of the axon terminal from entorhinal cortex layer 3.

Voltage and CSD spatial distribution were constant for all the animals (n = 7) and behavioral states, indicating that they are reflecting the underlying anatomy of inputs to the CA1 regions. The combination of CSD analysis and ICA decomposition of LFPs is revealed as a useful tool to precisely identify the different hippocampal layers and more importantly to separate and identify the current sources of the LFP. However, this would require more extensive analysis so the next step was to build a model of the rat hippocampus to get a better understanding of the relations between source geometry and voltage distribution in the structure.

The present results in freely moving animals confirm and extend our previous work in anesthetized rats [4, 18–20]. In the next paragraphs, it follows a brief summary of previous experimental findings closely related to what has been presented in this section.

In the urethane-anesthetized preparation the *radiatum* and *lm* components were also the main contributors to CA1 LFPs, however the CA1pyr showed significantly less relative power than in the awake animal. This can be explained by a reduced firing rate of CA1 perisomatic-targeting interneurons under urethane [27, 42]. With local pharmacological manipulations it was demonstrated that the radiatum component activity was selectively decreased by non-NMDA glutamate blockers (DNQX) [18]. Targeted blockade of the ipsilateral CA3 with lidocaine injections also selectively decreased *radiatum* IC power. Those tests confirmed that the synaptic glutamate currents evoked by the input of the Schaffer collateral pathway from CA3 to the str. radiatum dendrites of the CA1 pyramidal cells were the underlying cause of the LFP activity captured by the *radiatum* IC. For this reason, we also termed it as *Schaffer* component. Pharmacological manipulations rendered less clear results for the *lm* component. As well as the *radiatum* IC, *lm* component activity was impaired by local injection of DNQX, indicating its glutamatergic nature. However, it was also affected by GABAa blockers (bicuculline) pointing towards a contribution from inhibitory currents [4]. This can be explained by the activity of several types of interneurons, which target the CA1 pyramidal cell dendrites at the str. lacunosum-moleculare, including oriens-lacunosum and neurogliaform cells; most of them are strongly feedforward activated by the entorhinal layer 3 axons [3, 27–29].

To go beyond a mere characterization of the main sources of CA1 LFPs and to show the usefulness of studying pathway-specific LFP components instead of the original mixed LFP signal, these previous works focused on the analysis of the temporal dynamics of the *radiatum* component and its relation with CA1 and CA3 units. The low firing rate and functional clustering of CA3 pyramidal cells allowed us discriminating elementary synaptic events in the *radiatum* IC, which were termed as micro-field excitatory postsynaptic potentials (μ-fEPSPs; [18]). *Radiatum* IC activity shows an ordered succession of μ-fEPSPs that appear to be generated by functional clusters of CA3 pyramidal neurons, to which individual units are recruited variably. Such pattern implies a hierarchical internal operation of the CA3 region based on sequential activation of pyramidal cell assemblies. A fraction of these excitatory packets readily induces firing of CA1 pyramids and interneurons, the so-called Schaffer-driven spikes, revealing the synaptic origin in the output code of single units. This finding supports the postulate that synchronous activity in cell assemblies is a network language for internal neural representation [9, 23].

A subsequent work [19] assessed the plastic changes underwent in the CA3-CA1 pathway spontaneous activity following long-term potentiation (LTP) and determined how pairs of pre- and postsynaptic neurons modify spike transfer compared to the population. It was found that the ongoing *radiatum* IC activity and the share of postsynaptic spikes fired by Schaffer input specifically in CA1 units increases after LTP without significant change of the mean firing rate. A re-organization of the presynaptic cell assemblies synchronously firing to elicit CA1 spikes was also found. Thus, the results provided first time evidence for pathway-specific ongoing plasticity and its impact over spontaneous network activity consisting on the increased spike transfer between nuclei connected by specific potentiated channels. These results provide evidence that LTP induction produces a pathway-specific enhancement of ongoing activity that is effectively propagated to subsequent relays of the network. These observations complement and extend on classic LTP properties observed with evoked stimuli by showing their ongoing correlates and supports the view of synfire chains [1] as a prominent mechanism for information transfer in neural networks. Similar effects were also reported in awake animals after the administration of a pepide that activate the PI3K signaling pathway increasing spine density in the hippocampus and hippocampal-dependent learning [16].

3.1.2 DentateGyrus

Despite the fact that LFPs in dentate gyrus have been much less intensively researched than in the CA1 region, it has been know for long time that this structure exhibits a rich variety of LFP patterns and oscillations, including theta and gamma rhythms [5], dentate spikes [6], slow oscillations [26] and odor-evoked beta oscillations [24]. However, due to the complexity of its local circuits and the scarce knowledge regarding the synaptic inputs and firing properties of its different cell types during behavior, the mechanisms of generation of the different LFP patters observed in the DG remain largely unknown. It has been shown that DG theta and gamma oscillations are strongly modulated during exploratory and learning behavior in rodents [13, 21, 34, 40], pointing towards an important function of these rhythms in cognitive functions involving this structure. DG oscillatory dynamics also has a strong impact on its main target region, CA3, [2, 36, 37] and the computations performed in the whole hippocampal circuit [34, 41].

There are two main extrinsic afferences to the DG, the medial (MPP) and the lateral (LPP) perforant paths originating in layer 2 of medial and lateral entorhinal cortex and innervating the distal and middle thirds of granular cell (GC) dendrites. So is to be expected that these two inputs are major contributors to DG LFPs. However, there are many others inputs that can also contribute substantially. On one hand, the associational-commissural fibers innervate the inner third of GC dendrites and on the other the multitude of GC layer and hilar interneurons innervate the soma and dendritic regions of the GCs.

Following the same procedure as that previously described for the CA1 LFPs, we identify three main ICs in the DG. The three ICs have similar voltage loadings, with a plateau-like maximum between cell layers throughout the hilus, which declined outwardly and reversed its polarity at different points, and characteristics points for each of them (Fig. 3.3b). The CSD loading shows more differences between ICs.

The first IC (*LPP*, Fig. 3.3b; blue traces) is the one that reverses its polarity more superficially (closer to the fissure) and has a sink in the superficial GC dendrites and a source closer to their soma. This current distribution is similar to that obtained in the DG for evoked potentials stimulating the LPP [4, 31]. The second component (*MPP*, Fig. 3.3b; red traces) has a reversal point around 100 μm below the LPP and a sink in the middle third of the GC dendrites surrounded by two smaller sources at the distal dendrites and GC soma. This current distribution is similar to that obtained in the DG for evoked potentials stimulating the MPP [4, 20, 31]. The third IC (*GCsom*, Fig. 3.3b; green traces) has a source at the GC soma and a sink in the middle of the dendritic region and its voltage loading reverse at a similar depth than the *MPP* IC. This current distribution could be produced by a perisomatic inhibition, as it is the case of the dentate basket cells [22, 25].

The 2D plots for the three components further illustrate their characteristic spatial distributions. The voltage maps of the three of them are characterized by strong positive voltage in the hilus that decay toward the GC dendrites. In the three cases it reverts in the str. moleculare but only the *MPP* shows strong and localized negative voltage. This can be explained because the dipolar configuration of both *LPP* and GCsom favors a passive decay of the negative voltage gradient while the quadrupolar configuration of the *MPP* favors a more closed negative field. This is similar to what happened with the *radiatum* IC in CA1 which has also a quadrupolar configuration in contrast to the CA1 *lm* IC which has a dipolar one (Fig. 3.2b, c).

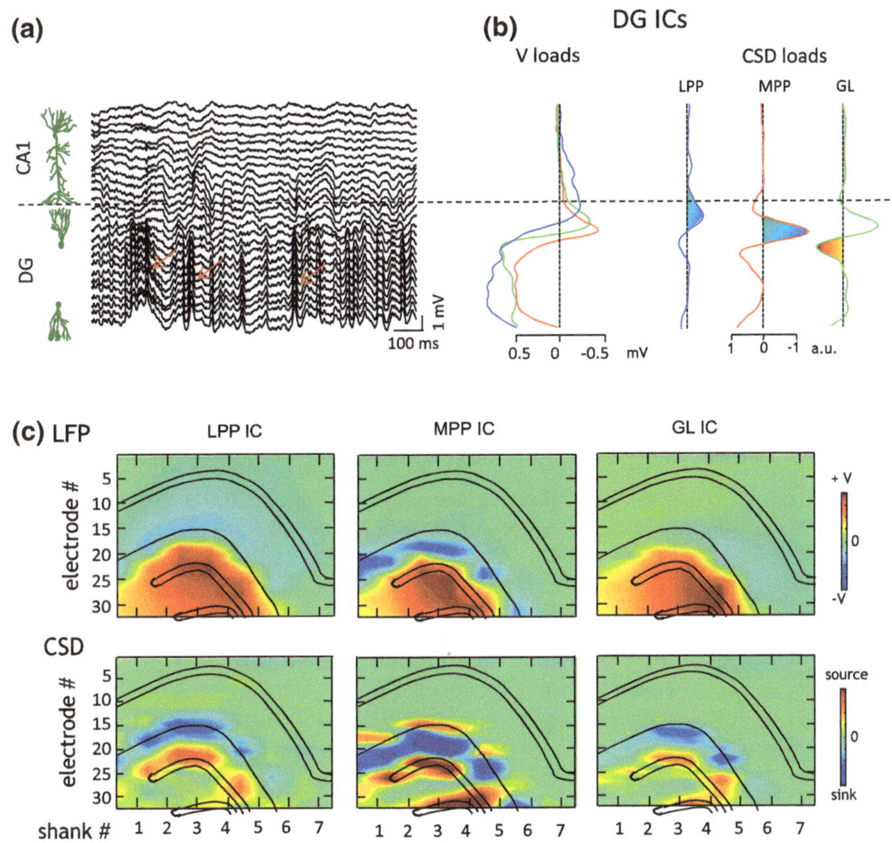

Fig. 3.3 **a** Similar LFP profile as illustrated in Fig. 3.2 but featuring a characteristic DG LFP pattern, dentate spikes (*red arrow*). **b** Three main ICs were found for DG LFPs. All of them display large positive voltage across the hilus but reverse polarity at different depths in the str. moleculare. Largest currents were restricted to the outer third of the str. moleculare (*LPP*), middle third (*MPP*) and GC layer (*GCsom*). **c** 2D voltage distributions for the three ICs were dominated for the positive hilar potentials but the CSD maps illustrated their different laminar specificity

The 2D CSD maps confirm the distal sink and inner dendritic source of the *LPP* IC, the source-sink-source configuration of the *MPP* and the dendritic sink-somatic source of the *GCsom* (Fig. 3.3c). It is important to note the lack of any currents for the three components in the hilus. That is to be expected as the axon is the only GC element in this region and it drains only a negligible amount of current.

As was also the case for the CA1 LFP ICs, the present results for the DG LFPs confirm and extend our previous results in urethane-anesthetized rats [4, 20]. In those previous works, it was shown with local pharmacological manipulations that the activity of the *MPP* and *LPP* components was selectively decreased with the injection of DNQX (a non-NMDA glutamate blocker). On the contrary, the *GCsom* component was affected by both glutamate and GABAa (bicuculline) blockers.

This result can be explained by two mechanisms. Either the activity captured by the *GCsom* component is contributed by excitatory and inhibitory currents onto the perisomatic region of the GC or it is only elicited by inhibitory currents delivered by interneurons relaying for its activation on excitatory inputs. The lack of somatic excitatory inputs leads us to think in the second alternative as the most plausible.

Another test for the pathway-specificity of the LFP ICs was achieved in those previous works by the electrical stimulation of afferent pathways to the hippocampus [4, 18, 20]. When subthreshold stimuli (not strong enough to evoked a population spike, i.e., the synchronous discharge of action potentials in the target population) were delivered to the medial and lateral perforant paths and the Schaffer collaterals, a field EPSP (fEPSP) was recorded in the DG and CA1 respectively, indicating synchronous excitatory synaptic currents onto the target populations of cells [30, 31]. After ICA decomposition of the LFPs, those fEPSPs were captured by the *MPP*, *LPP* and *radiatum* components selectively. This result confirms that the currents elicited by both the spontaneous and evoked activity of those pathways are selectively captured by specific ICs. Thus, ICA decomposition of hippocampal LFPs allows the separation and identification of pathway-specific contributors to the LFPs [17].

3.1.3 Characterization of Hippocampal LFP Sources

The temporal and spectral dynamics of the LFP ICs during different behavioral states contain a large amount of information about the computations performed by the hippocampal circuits [39]. Although I will extend on this aspect in the last section of the Results, some gross quantifications of the ICs activity are provided in Fig. 3.4.

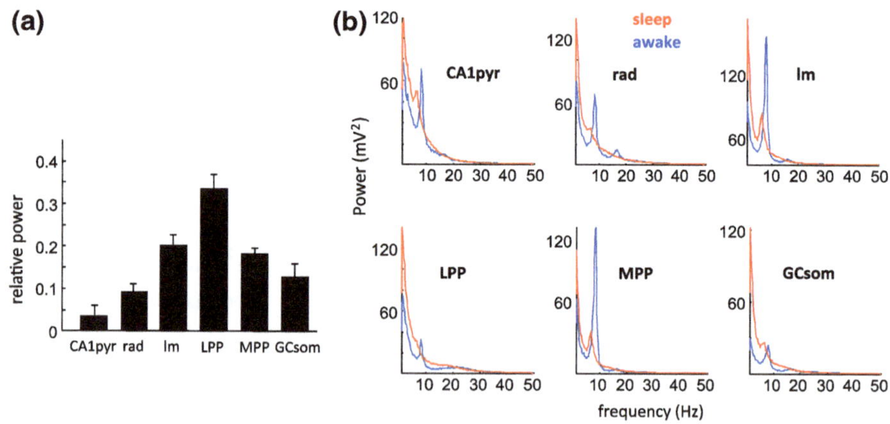

Fig. 3.4 **a** Relative power for the six main hippocampal LFP ICs averaged across shanks, sessions and animals (n = 7 animals). **b** Power spectrum of the six ICs for periods where animals were sleeping in their home cages (*red*) or performing different navigational tasks (*blue*). Data averaged across shanks, sessions and animals (n = 7 animals)

To quantify the contribution of each IC to the recorded LFPs we calculated the relative variance of the LFP matrix that is accounted by each IC. The results in Fig. 3.4a were obtained pooling together all the ICs extracted during sleep and behavior in all the recording shanks that cover both CA1 and DG regions in seven animals. The 6 main ICs reported here usually account for more than 95 % of the total variance of the LFPs in the selected recording tracks, once the noise and artifactual activity was removed (see Sect. 3.1.3). The remaining variance was explained by other sources with very small contribution to the LFP.

The IC with largest variance was the *LPP*, followed by the *lm* and *MPP* components. Together the three DG ICs have more variance that the three CA1 ICs, as is expected due to fact that DG LFPs are in general larger that CA1 LFPs [20]. In the second chapter of the Results, I will analyze the biophysical factors that explain this phenomenon. The IC with lower variance was the *CA1pyr*, what can be explained by having its current restricted to somatic region. An important factor determining the contribution of the different current sources to the LFPs is their dipolar moments. Sources with a larger dipolar moment, as the *lm* and *LPP*, which have the geometrical configuration of a linear dipole are expected to produce larger fields [32, 38]. On the contrary, sources like *rad*, *CA1pyr* or *MPP* ICs have the geometry of a linear quadrupole, thus a shorter dipolar moment, and are expected to produce closer fields and contribute less to the LFPs.

In Fig. 3.4b averaged power spectrums of the six ICs corresponding to times while rats are sleeping in his home cage (red traces) or awake performing different navigational tasks (blue traces) are compared. Although each IC has different spectral dynamics the averaged spectrums during behavior are dominated by the ubiquitous presence of the theta rhythm (note the peak around 8 Hz). The spectral theta peak is more prominent in the *lm* component, as corresponds to the largest theta dipole being located in the str. lacunosum-moleculare [7, 8]. During sleep there are two differentiated stages, the slow-wave sleep characterized by 1–2 Hz oscillations, and the rapid-eye movement sleep, characterized by theta oscillations of slightly lower frequency than those present during locomotion. Both oscillations are visible with different relative power in the spectrums of the six ICs.

In this section we have shown that the application of ICA to large-scale recordings of hippocampal LFPs is able to disentangle their underlying synaptic sources. We found three main sources of CA1 and DG LFPs respectively. Each of them was characterized by a restricted laminar distribution of currents that allows their matching with the known synaptic domain of main extrinsic and intrinsic inputs to the CA1 pyramidal cells and DG granular cells populations. In addition, those sources display different spectral characteristics. This suggests the possibility that the study of their spectro-temporal dynamics would be informative to understand the computations performed by the hippocampal circuits during behavior. This will be extensively explored in following sections.

3.2 Finite Elements Simulations of LFP's

The combination of ICA and CSD analysis allowed us to solve the inverse problem of the LFPs; that is, given the recorded LFPs, separate and identify their underlying current sources. As a test of the accuracy of the above results, we sought to use the extracted LFP sources to solve the forward problem of the LFPs, i.e., to reproduce the original LFP distribution. For this purpose, we built a 3D model of the dorsal hippocampus of the rat with FEM (see Methods).

The model simulates a transversal lamella of the dorsal hippocampus similar to that were the recording electrode was placed in all the animals (Fig. 3.1b). This lamella was extended 4 mm in depth to simulate a whole block of the dorsal hippocampus. All the simulated recordings were performed in the middle of the structure to minimize any possible border effect.

In the model each hippocampal layer was represented as a polygonal block. Thus, the cellular and dendritic portions were represented as stacked longitudinal blocks of current sources, each representing a subcellular "population" domain that when activated act as laminar dipoles. In preliminary simulations we found that eighth of these blocks with 100 μm thickness in the case of CA1 and four 60 μm thickness blocks for each blade of the DGwere sufficient to reproduce the main laminar features of hippocampal LFPs. To implement the exact geometry of the current sources of the LFP, average CSD loadings from each LFPs ICs were discretized into eight or four point curves that were used as weights to implement the sources in CA1 and DG region of the model (see Methods).

In the first set of simulations, we implement in the model the three main sources of CA1 and DG LFPs that were identified in the previous section in isolation. We took 100 s of activity from the six main ICs extracted while the rat was resting in the home cage, and use them as inputs to the whole CA1 or DG regions, that were assumed to be simultaneously and homogeneously active.

LFPs were sampled in a linear track in the middle of the hippocampus, approximately perpendicular to the cellular layers (Figs. 3.5a and 3.6a). After the independent activation of each source, the averaged voltage power along this line was calculated and used to construct voltage loading curves for each IC and its second derivative to construct CSD loading curves (Figs. 3.5b and 3.6b).

For CA1 LFP ICs, voltage and CSD loadings obtained from the activation of the three sources were remarkably similar to those obtained experimentally (compare Figs. 3.3b and 3.2b). The *CA1pyr* component displays a sharp peak in its voltage loading at the pyramidal layer and the corresponding source surrounded by two sinks in the CSD loading (Fig. 3.5b; blue traces). The largest amplitude of the *rad* IC is at the level of the *str. radiatum*, around 200 μm below the pyramidal layer, and for the *lm* IC at the str. lacunosum-moleculare, around 500 μm below the pyramidal layer. The *rad* component CSD loading has a larger sink flanked by two smaller sources while the lm displays a rather symmetrical current dipole (Fig. 3.5b; red and green traces).

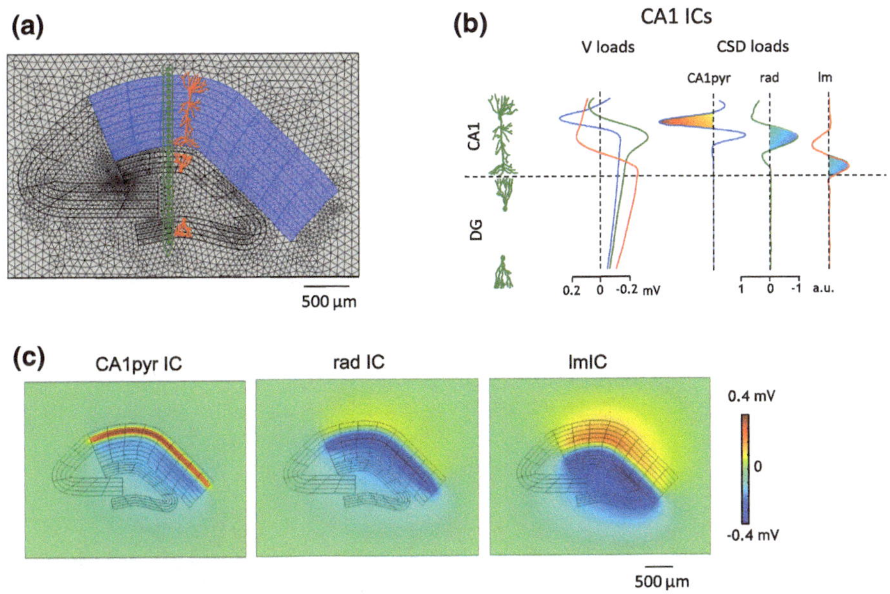

Fig. 3.5 **a** FEM model of the rat dorsal *hippocampus* highlighting the CA1 region (*blue*). **b** Voltage and CSD loadings along a vertical profile spanning CA1 and DG (indicated by the position of the *green linear* probe in **a**) illustrating the simulated CA1 LFP sources (note their similarity with the experimentally obtained ICs in Fig. 3.2b). **c** Averaged 2D voltage distributions obtained with the activation of the 3 CA1 LFP sources illustrate their layer-specific distribution

The model also allows us to calculate the voltage distribution in the whole hippocampus and surrounding extracellular space. We took a 2D plane parallel to the transversal axis of the hippocampus situated in the middle of the structure and plot the voltage distribution resulting from the activation of the three sources independently (Fig. 3.3c). The 2D maps obtained were highly similar to the pseudo 2D voltage maps constructed from the voltage loadings of the ICs obtained experimentally (compare Figs. 3.5c and 3.2b). In the three cases, the laminar distribution of the three ICs matches with the corresponding CA1 sublayers.

Current source 2D maps were omitted due to the fact that they did not offer any additional information given that this was exactly what was introduced in the model initially.

We repeat the above procedure for the DG ICs. As for the ICs extracted from the recorded LFPs, the three main DG ICs have voltage loadings with large positive amplitudes across the whole hilus which decay beyond the GC layers and reverse at different depths in the str. moleculare (compare Figs. 3.6b and 3.3b). The *LPP* component has the most superficial reversal point, as correspond to its sink in the outer third of the str. moleculare (Fig. 3.6b; blue traces). The *MPP* IC reverses closer to the GC layer and displays a large sink in the middle of the str. moleculare flanked by two smaller sources (Fig. 3.6b; blue traces). The *GCsom* has a voltage

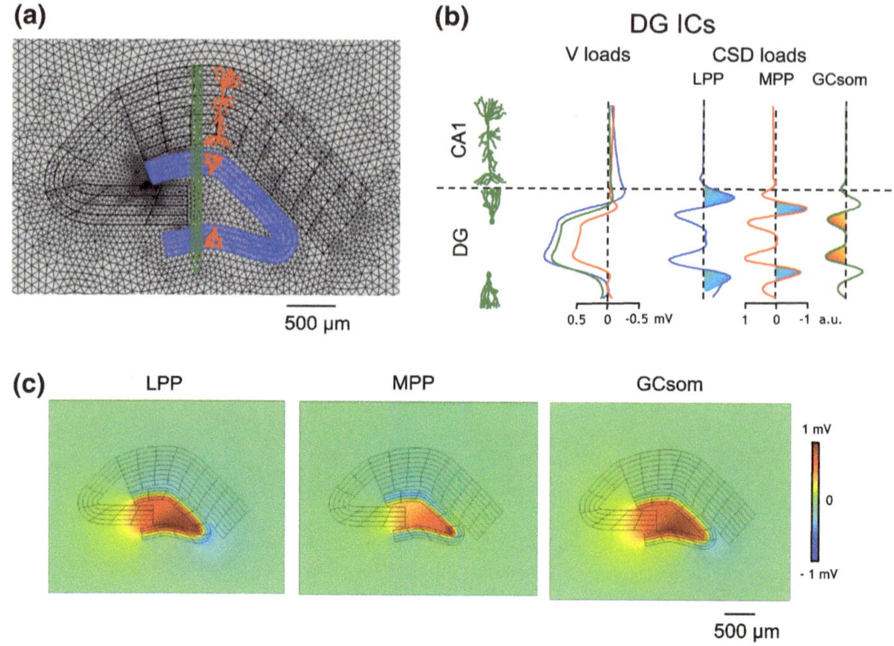

Fig. 3.6 **a** FEM model of the rat dorsal *hippocampus* highlighting the DG region (*blue*). **b** Voltage and CSD loadings along the same vertical profile as in Fig. 3.5 of the three simulated DG LFP sources (note their similarity with the experimentally obtained ICs in Fig. 3.3b). **c** Averaged 2D voltage distributions obtained with the activation of the 3 DG LFP sources are dominated by positive hilar potentials but display different location and spread of negative potential in dendritic layers

and CSD loading with reversal point and source location intermediate between the *LPP* and *MPP* ones.

2D voltage maps for the three DG ICs display also similar features with those constructed from the experimental data (compare Figs. 3.6c and 3.3c), in all the cases dominated by large hilar positive potentials. Note the more restricted negative voltage of the *MPP* ICs compared to the *LPP*; this can be explained by the more "close-field" configuration of the *MPP*, as will be analyzed in the next section. In comparison the *GCsom* IC elicits very small negative fields.

The above results show that the presented model accurately reproduces the voltage distributions experimentally obtained for the six main current sources of CA1 and DG LFPs. However, a question remains about if those sources when activated simultaneously with similar dynamics as those observed in the experiments reproduce the observed profile of LFPs. To answer this question we performed another set of simulations in which the six sources were activated simultaneously.

We performed these simulations feeding the model with the ICs time series extracted in either during sleep or during running activity. By doing this, we sought to reproduce in the model the two main hippocampal LFP patterns: theta oscillations and sharp-wave ripples, SWR (see Introduction). To sample LFPs a recording linear probe with electrodes spaced 50 μm was simulated in the middle of the hippocampus, spanning the CA1 and DG regions.

SWR are characterized by a large negative LFP wave, and the corresponding sink, at the CA1 str. radiatum which reverse at the pyramidal layer, and it is accompanied by sources at the pyramidal layer and str. lacunosum-moleculare (Fig. 3.7a). This activity is caused by a synchronous input of CA3 pyramidal cells to CA1 eliciting a large depolarization of str. radiatum dendrites accompanied by their return passive currents [5, 6, 18, 19]. The averaged power along the CA1-DG during SWR is dominated by a large increase at the str. radiatum (Fig. 3.7a; histogram). In the simulated LFPs all the main characteristics of the LFP and CSD profiles of SWR were replicated, remarkably the large negative LFP amplitude at the str. radiatum and the characteristic source-sink-source current distribution (Fig. 3.7b).

The CSD map of theta oscillation is more complex that the one obtained for SWR because in the former case is not only one pathway (i.e., source) which is

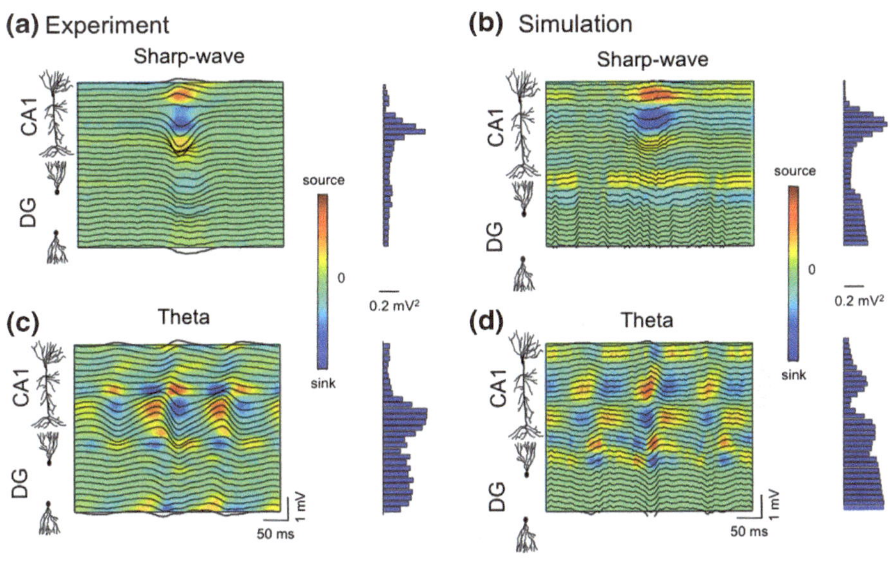

Fig. 3.7 **a** Averaged LFP profile of sharp-wave ripples show a large negative LFP deflection at str. radiatum and the corresponding dominating power at that layer (*histogram*). The CSD map illustrates the characteristic source-sink-source distribution for the excitatory Schaffer input to the CA1 pyramidal cells dendrites. **b** LFP, CSD and power profiles were constructed in the same way for simulated SWR and display similar laminar profile as the experimental ones. **c** Averaged LFP and CSD profile for theta oscillation during running. Largest currents and LFP power were present in the str. lacunosum-moleculare but phase shifted sink and sources also appear at str. radiatum and DG str. moleculare. **d** Simultaneous activation of the six hippocampal LFP sources presented in this section with temporal activations extracted from experimental recordings during running result in a similar laminar potential and CSD distribution

dominating but the simultaneous activity of several, if not all, of them [7, 8]. The largest theta currents and fields are present at the str. lacunosum-moleculare but they are accompanied by shifted sink/sources at the str. radiatum and str. moleculare of the DG and with lower intensity in str. pyramidale and oriens (Fig. 3.7c). Along the depth profile theta waves gradually shift their polarity displaying a 180° reversal between pyramidal layer and str. lacunosum-moleculare. Both the LFP and CSD profile characteristics of theta oscillations were replicated in our simulations.

With the above simulations we were able to prove that our model accurately reproduces the laminar characteristics of hippocampal LFPs. This confirms that the six current sources identified with ICA are sufficient to account for the main LFP patterns observed in the hippocampus. Having settled the fundamental basis of our analytical approach, in the next sections we will employ the FEM hippocampal model and the ICA + CSD decomposition of LFPs to investigate the biophysical and physiological mechanisms of LFP generation in the hippocampus as a tool to interpret the underlying activity of neural circuits.

References

1. Abeles M (1991) Corticonics. Cambridge, Cambridge University Press
2. Akam T, Oren I, Mantoan L, Ferenczi E, Kullmann DM (2012) Oscillatory dynamics in the hippocampus support dentate gyrus–CA3 coupling. Nat Neurosci 15(5):763–768
3. Basu J, Srinivas KV, Cheung SK, Taniguchi H, Huang ZJ, Siegelbaum SA (2013) A cortico-hippocampal learning rule shapes inhibitory microcircuit activity to enhance hippocampal information flow. Neuron 79:1208–1221
4. Benito N, Fernandez-Ruiz A, Makarov VA, Makarova J, Korovaichuk A, Herreras O (2014) Spatial modules of coherent activity in pathway-specific LFPs in the hippocampus reflect topology and different modes of presynaptic synchronization. Cereb Cortex 24:1738–1752
5. Bragin A, Jandó G, Nádasdy Z, Hetke J, Wise K, Buzsáki G (1995) Gamma (40-100 Hz) oscillation in the hippocampus of the behaving rat. J Neurosci 15:47–60
6. Bragin A, Jandó G, Nádasdy Z, van Landeghem M, Buzsáki G (1995) Dentate EEG spikes and associated interneuronal population bursts in the hippocampal hilar region of the rat. J Neurophysiol 73:1691–1705
7. Brankack J, Stewart M, Fox SE (1993) Current source density analysis of the hippocampal theta rhythm: associated sustained potentials and candidate synaptic generators. Brain Res 615:310–327
8. Buzsáki G (2002) Theta oscillations in the hippocampus. Neuron 33:325–340
9. Buzsáki G (2010) Neural syntax: cell assemblies, synapsembles, and readers. Neuron 68: 362–385
10. Buzsáki G, Anastassiou CA, Koch C (2012) The origin of extracellular fields and currents—EEG, ECoG, LFP and spikes. Nat Rev Neurosci 13:407–420
11. Buzsáki G, Czopf J, Kondakor I, Kellenyi L (1986) Laminar distribution of hippocampal rhythmic slow activity (RSA) in the behaving rat: current source density analysis, effects of urethane and atropine. Brain Res 365:125–137
12. Buzsáki G, Leung LS, Vanderwolf CH (1983) Cellular bases of hippocampal EEG in the behaving rat. Brain Res 287:139–171
13. Csicsvari J, Jamieson B, Wise KD, Buzsáki G (2003) Mechanisms of gamma oscillations in the hippocampus of the behaving rat. Neuron 37:311–322

14. de Cheveigné A, Edeline JM, Gaucher Q, Gourévitch B (2013) Component analysis reveals sharp tuning of the local field potential in the guinea pig auditory cortex. J Neurophysiol 109: 261–272
15. Einevoll GT, Kayser C, Logothetis NK, Panzeri S (2013) Modelling and analysis of local field potentials for studying the function of cortical circuits. Nat Rev Neurosci 14:770–785
16. Enriquez-Barreto L, Cuesto G, Dominguez-Iturza N, Gavilán E, Ruano D, Sandi C, Fernández-Ruiz A, Martín-Vázquez G, Herreras O, Morales M (2014) Learning improvement after PI3K activation correlates with de novo formation of functional small spines. Front Mol Neurosci 2(6):54
17. Fernández-Ruiz A, Herreras O (2013) Identifying the synaptic origin of ongoing neuronal oscillations through spatial discrimination of electric fields. Front Comput Neurosci 7:5
18. Fernández-Ruiz A, Makarov VA, Benito N, Herreras O (2012) Schaffer-specific local field potentials reflect discrete excitatory events at gamma frequency that may fire postsynaptic hippocampal CA1 units. J Neurosci 32:5165–5176
19. Fernández-Ruiz A, Makarov VA, Herreras O (2012) Sustained increase of spontaneous input and spike transfer in the CA3-CA1 pathway following long-term potentiation in vivo. Front Neural Circ 6:71
20. Fernández-Ruiz A, Muñoz S, Sancho M, Makarova J, Makarov VA, Herreras O (2013) Cytoarchitectonic and dynamic origins of giant positive local field potentials in the dentate Gyrus. J Neurosci 33:15518–15532
21. Givens B (1996) Stimulus-evoked resetting of the dentate theta rhythm: relation to working memory. NeuroReport 8:159–163
22. Han ZS, Buhl EH, Lörinczi Z, Somogyi P (1993) A high degree of spatial selectivity in the axonal and dendritic domains of physiologically identified local-circuit neurons in the dentate gyrus of the rat hippocampus. Eur J Neurosci 5:395–410
23. Harris KD (2005) Neural signatures of cell assembly organization. Nat Rev Neurosci 6: 399–407
24. Heale VR, Vanderwolf CH, Kavaliers M (1994) Components of weasel and fox odors elicit fast wave bursts in the dentate gyrus of rats. Behav Brain Res 63:159–165
25. Houser CR (2007) Interneurons of the dentate gyrus: an overview of cell types, terminal fields and neurochemical identity. Prog Brain Res 163:217–232
26. Isomura Y, Sirota A, Ozen S, Montgomery S, Mizuseki K, Henze DA, Buzsáki G (2006) Integration and segregation of activity in entorhinal-hippocampal subregions by neocortical slow oscillations. Neuron 52:871–882
27. Klausberger T, Somogyi P (2008) Neuronal diversity and temporal dynamics: the unity of hippocampal circuit operations. Science 321:53–57
28. Lasztóczi B, Klausberger T (2014) Layer-specific GABAergic control of distinct gamma oscillations in the CA1 hippocampus. Neuron 81:1126–1139
29. Leão RN, Mikulovic S, Leão KE, Munguba H, Gezelius H, Enjin A, Patra K, Eriksson A, Loew LM, Tort ABL, Kullander K (2012) OLM interneurons differentially modulate CA3 and entorhinal inputs to hippocampal CA1 neurons. Nat Neurosci 15:1524–1530
30. Leung LS (1984) Model of gradual phase shift of theta rhythm in the rat. J Neurophysiol 52:1051–1065
31. Leung LS, Roth L, Canning KJ (1995) Entorhinal inputs to hippocampal CA1 and dentate gyrus in the rat: a current-source-density study. J Neurophysiol 73:2392–2403
32. Malmivuo J, Plonsey R (1995) Bioelectromagnetism: principles and applications of bioelectric and biomagnetic fields. Oxford University Press, USA
33. Mizuseki K, Diba K, Pastalkova E, Buzsáki G (2011) Hippocampal CA1 pyramidal cells form functionally distinct sublayers. Nat Neurosci 14:1174–1181
34. Montgomery SM, Betancur MI, Buzsáki G (2009) Behavior-dependent coordination of multiple theta dipoles in the hippocampus. J Neurosci 29:1381–1394
35. Montgomery SM, Buzsàki G (2007) Gamma oscillations dynamically couple hippocampal CA3 and CA1 regions during memory task performance. Proc Natl Acad Sci USA 104: 14495–14500

36. Mori M, Abegg MH, Gähwiler BH, Gerber U (2004) A frequency-dependent switch from inhibition to excitation in a hippocampal unitary circuit. Nature 431:453–456
37. Neunuebel JP, Knierim JJ (2014) CA3 retrieves coherent representations from degraded input: direct evidence for CA3 pattern completion and dentate gyrus pattern separation. Neuron 81:416–427
38. Nunez PL, Srinivasan R (2006) Electric fields of the brain: the neurophysics of EEG, 2nd edn. Oxford University Press, New York
39. Schomburg EW, Fernández-Ruiz A, Berényi A, Mizuseki K, Anastassiou CA, Koch C, Buzsáki G (2014) Theta phase segregation of input-specific gamma patterns in entorhinal-hippocampal networks. Neuron 84:470–485
40. Skaggs WE, McNaughton BL, Wilson MA, Barnes CA (1996) Theta phase precession in hippocampal neuronal populations and the compression of temporal sequences. Hippocampus 6:149–172
41. Sullivan D, Csicsvari J, Mizuseki K, Montgomery S, Diba K, Buzsáki G (2011) Relationships between hippocampal sharp waves, ripples, and fast gamma oscillation: influence of dentate and entorhinal cortical activity. J Neruosci 31:8605–8616
42. Ylinen A, Bragin A, Nádasdy Z, Jando G, Szabó I, Sik A, Buzsáki G (1995) Sharp wave associated high-frequency oscillation (200 Hz) in the intact hippocampus: network and intracellular mechanisms. J Neurosci 15:30–46

Chapter 4
Theta-Gamma Cross-Frequency Coupling in the Hippocampus-Entorhinal Circuit

The hippocampal-entorhinal system is characterized by the ubiquitous occurrence of distinct oscillatory patterns, including the prominent theta and gamma rhythms [9, 12, 21, 32, 47]. Theta phase coordination of gamma rhythms within and across brain regions has been studied extensively, however its cellular mechanism and functional relevance remain largely unknown [6, 10, 16, 17, 21, 23, 33, 34, 44].

The CA1 region of the hippocampus is under the control of two major upstream regions: hippocampal area CA3 and the entorhinal cortex. CA3 axons make synapses with CA1 pyramidal cells in the stratum radiatum while layer 3 entorhinal cells (EC3) innervate the stratum lacunosum-moleculare [2, 48]. Their layer-segregated inputs mediate both dendritic excitation and feedforward inhibition in CA1 [8]. To determine the coordination of gamma oscillations by the theta rhythm in CA1, the dynamic interactions between the entorhinal and CA3 inputs to the CA1 region, and their impact on the CA1 output, we used high-density extracellular recordings, combined with source separation techniques, while rats performed different navigational tasks and slept in their home cages.

Experiments were carried out while animals ran on a linear track (250 cm long), a T-maze or open field [20, 35, 37, 38, 40, 43]. Theta epochs during behavioral tasks were classified as RUN, while those during sleep were classified as REM.

4.1 Sources of Gamma Oscillations in CA1

It was already known that the phase of theta rhythm modulates the power of gamma oscillations; however, large controversy remains regarding how many independent gamma oscillators are present in the CA1 region and how are their exact interactions [4, 14, 17, 31, 41]. Thus the first goal of this part of our research was to clarify those points.

We took the times of theta activity during both sleep (REM) and behavior (RUN) and concatenate epochs until having acceptable homogeneous samples of

A. Fernández-Ruiz, *Extracellular Potentials in the Hippocampus*, Springer Theses, DOI 10.1007/978-3-319-41039-5_4

600 s that were used for subsequent analyses. RUN/REM comparisons were performed for theta epochs from the same session and animal.

We filtered CA1 pyramidal layer LFP in a broad gamma band, from 30 to 300 Hz, and extracted the signal amplitude for every frequency step by means of complex wavelet transform (see Methods). We also filtered the LFP between 1 and 20 Hz and by the same procedure extracted the phase for every frequency step. Then we calculated the modulation of the high frequency amplitude by the phase of the slower frequency employing the modulation index (MI) introduced by [46]. With the MI for every amplitude-phase pairs we constructed 2D comodulogram plots (Fig. 4.1a, b). This procedure reveals three distinct but overlapping gamma sub-bands during RUN (Fig. 4.1a, first panel). One of the oscillations occupies the slower gamma spectrum (30–60 Hz) so I labeled it as slow gamma or $gamma_S$, another gamma oscillations span from 60 to 100 Hz so I termed it as mid-frequency gamma or $gamma_M$. The third component spans a wide frequency range of the upper gamma spectrum, from 120 to 250 Hz, so I named it as fast gamma or $gamma_F$.

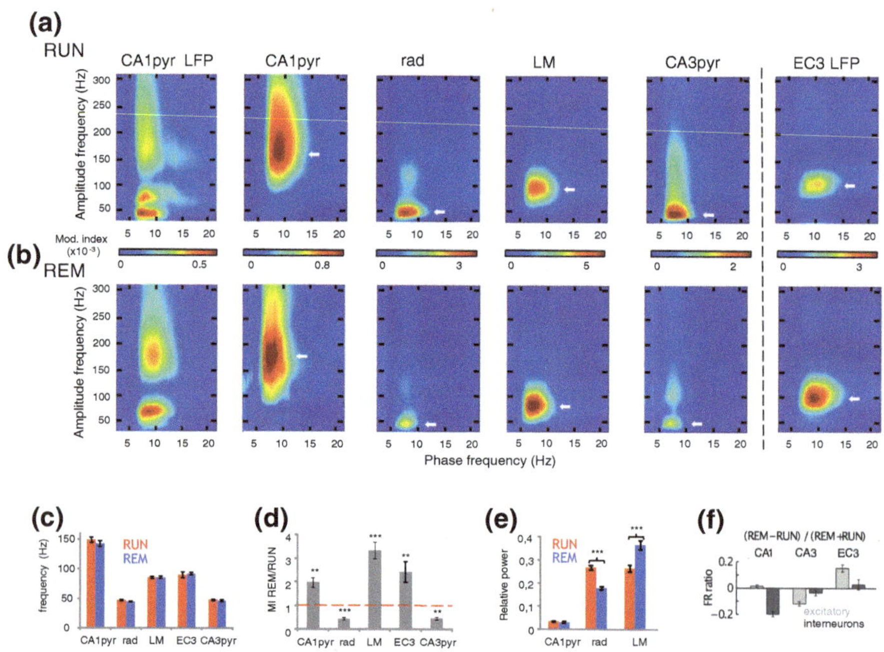

Fig. 4.1 **a** Gamma amplitude-theta frequency comodulogram of LFP in the CA1 pyramidal layer (CA1 str. pyr. LFP) showed strong theta phase modulation of three gamma sub-bands during RUN ($gamma_S$, $gamma_M$, and $gamma_F$; white arrows). Each IC obtained for CA1 LFPs displayed modulation in one dominant sub-band. White arrows indicate the frequency of peak power. Note similar gamma frequencies in the rad and CA3pyr ICs, and the lm IC and EC3 LFP, respectively. **b** Same as in **a** during REM sleep. **c** Group data of peak frequencies (six rats for CA1 and CA3 ICs, four rats for EC3 LFP). **d** and **e** REM/RUN MI ratio (D) and relative power (30–300 Hz) in different layers (**e**). (*/**/***$p < 0.05/0.01/0.001$; t test). **f** Mean ± SEM of firing rate (FR) ratios of single units between REM and RUN ([REM–RUN]/[REM + RUN]). Reproduced from [43]

Because underlying currents of gamma LFP in the pyramidal layer reflect a combination (in unknown proportions) of active excitatory and inhibitory currents and passive return currents from the dendritic layers [19, 22, 27, 42], we employed ICA to separate the observed gamma oscillations of the raw LFP into their pathway-specific synaptic sources (see Methods). For this purpose we took the LFPs from all the electrodes located in the CA1 region (thus including the pyramidal and dendritic layers), filtered them between 30 and 300 Hz and performed ICA. The same three main ICs as shown in the previous chapter and previous work were found: *CA1pyr*, *radiatum* (*rad*) and *lac-mol* (*lm*) [5, 24]. Cross-frequency phase-amplitude analysis of the three ICs revealed significantly theta-modulated gamma bands in all animals ($p < 0.001$ for each IC, surrogate test; 7 animals in total). In the *CA1pyr* IC, theta oscillations most strongly modulated fast gamma frequencies (Fig. 4.1a, second panel; mean ± s.e.m., 149.4 ± 4.3 Hz). In the *rad* IC, the dominant theta-modulated gamma frequencies were between 30 and 70 Hz (Fig. 4.1a, third panel; 47.3 ± 0.6 Hz). Compared to the *rad* IC, theta-coupled gamma oscillations in the *lm* IC were significantly faster ($p < 0.0001$, t-test; Fig. 4.1a, fourth panel; 85.7 ± 1.8 Hz). Thus ICA decomposition of CA1 gamma LFPs was able to separate the contribution of three independent gamma generators, each one with activity in a discrete frequency band.

To clarify the origin of those gamma generators we looked at the CA1 input regions. CA3 pyramidal cells send their axons to the CA1 stratum radiatum and it was already now that they elicit strong slow gamma there [19, 24, 25, 50], so they are the most likely candidates for being responsible of radiatum gamma$_S$. When we applied the same procedure as described in this section to the LFP recorded in the CA3 pyramidal layer, we found a theta-modulated gamma band similar to that in the *rad* IC ($p > 0.05$, t-test between frequencies), with a peak frequency of 47.6 ± 1.2 Hz (Fig. 4.1a, fifth panel). These similar gamma oscillations can be attributed to the fact that the same CA3 pyramidal cells that send axons to the CA1 also leave collaterals in the same CA3 region (Li et al. 1994; Ishikuza et al. 1990). The CA1 stratum lacunosum-moleculare is densely innervated by axons from the entorhinal cortex layer 3 so we also applied the same analysis to the LFPs recorded there. Gamma oscillations there, were similar to those displayed by the *lm* IC ($p > 0.05$, t-test between frequencies) with a peak frequency of 90.0 ± 4.9 Hz (Fig. 4.1a, sixth panel), pointing also to EC3 projecting cells as the responsible for the *lm* gamma$_M$. The *CA1pyr* IC is most likely of local origin because the pyramidal layer and perisomatic region do not receive extrinsic afferences as the dendritic layers but is innervated by multitude of CA1 interneurons. Its peak frequency around 150 Hz suggests that it can be elicited by fast $GABA_A$ IPSPs onto the CA1 pyramidal cells but the fact that its theta-modulated gamma frequency extends up to 250 Hz, suggests that it can capture also contributions from action potentials [29, 41, 42].

REM sleep is characterized by prominent theta rhythm in the hippocampus so we sought to verify if theta-gamma dynamics described for the RUN state were preserved during sleep. We found the same three ICs for CA1 gamma LFPs during sleep. The frequency distribution of the theta modulated gamma sub-bands was largely similar to waking for all of them and also for CA3pyr and EC3 LFPs (Fig. 4.1b, c; $p > 0.05$, t-test). We employed two different methods to quantify the differences between both states. The modulation index (MI, Tort et al. 2008) to quantify the strength of theta-gamma coupling and the relative power (or relative variance of the IC; [26] of the gamma filtered ICs. Both MI and power of gamma$_S$ in the *rad* IC were significantly reduced during REM compared to RUN ($p < 0.0001$, t-test; Fig. 4.1b, d, e), whereas theta-gamma$_M$ coupling and power in the *lm* IC were significantly increased ($p < 0.0001$, t-test; Fig. 4.1b, d, e). These changes were accompanied by a parallel reduction of theta-gamma$_S$ coupling in CA3pyr ($p < 0.01$, t-test; Fig. 4.1b, d) and increased theta-gamma$_M$ coupling in EC3 LFP during REM ($p < 0.01$, t-test; Fig. 4.1b, d).

We also checked the firing of the neurons in the CA1, CA3 and EC3 to see if their changes in firing rate matched with those observed in the LFPs during waking and sleep. Individual units were extracted from the high-pass filtered LFPs and were classified into putative pyramidal cells and interneurons based on their waveform, autocorrelograms and other characteristics (see Methods). For each individual neuron the mean firing rate was calculated for both states and a (REM − RUN)/(REM + RUN) ratio used for comparison. Consistent with the LFP changes, CA3 and EC3 pyramidal neuron firing rates decreased and increased, respectively, during REM compared to RUN ($p < 0.05$, Kruskal-Wallis ANOVA, followed by Tukey's honestly significant difference test; Fig. 4.1f). In summary, theta-modulated gamma power in the respective dendritic domains of CA1 pyramidal cells mainly reflects the gamma band activity in their respective afferent regions and is modulated as a function of brain state and network architecture.

4.2 Coherence Segregation of Layer-Specific Gamma Sources

If gamma oscillations in the hippocampal CA1 are layer specific as I proposed in the previous section, is to be expected that their within-layer coherence is much larger than their cross-layer coherence. As a first step to verify this, coherence maps in the broad gamma frequency band (30–100 Hz) were constructed between LFPs at reference sites in different layers and the remaining 255 channels. This procedure reliably outlined the anatomical boundaries in CA1 for stratum pyramidale (Fig. 4.2a, first panel), stratum radiatum (second panel), and stratum lacunosum-moleculare (third panel).

Then we compared the gamma coherence (30–100 Hz) for all the three main CA1 LFP ICs extracted in every shank of the 8-shanks probe were the appropriated layer was recorded. The coherence matrix for a single case is displayed in Fig. 4.2b. High coherence values (warm colors) were only obtained for same ICs in separate shanks but not across shanks. The across animals quantification of coherence relative to distance (n = 6 rats; Fig. 4.2c) shows similar results. Gamma coherence

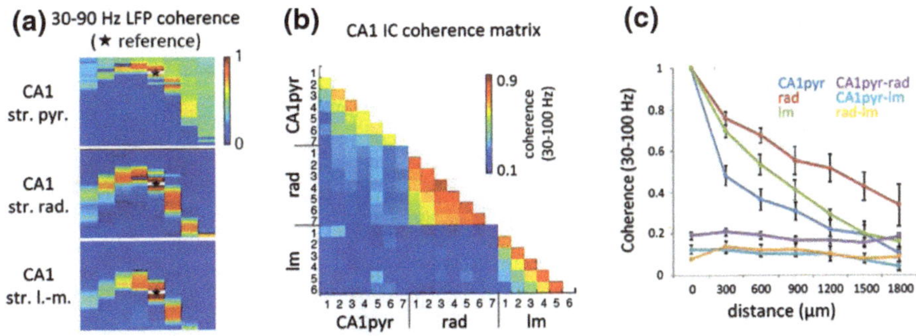

Fig. 4.2 **a** Gamma (30–90 Hz) coherence maps between LFP recorded from a reference site (white patch with star) and every other recording site on a 256-channel probe spanning most of a transverse plane in the dorsal hippocampus in one example session, for CA1 str. pyramidale (top), str. radiatum (middle), and str. lacunosum-moleculare (bottom) references. **b** Gamma coherence between pathway-specific CA1 ICs (extracted separately for each shank). Similar to cross-layer LFP coherence, ICs reflecting different synaptic pathways exhibited low coherence with other CA1 ICs across all shanks (numbered 1–7), but high coherence between like ICs from different shanks. **c** Coherence of gamma ICs decreased monotonically with distance between shanks, whereas coherence between different ICs was low, regardless of shank separation. Modified from [43]

remained relatively high (>0.4) for ICs from different shanks in the same layer, even up to 1.8 mm away, whereas gamma coherence between ICs from different layers was consistently lower even in the same shank.

4.3 Theta-Phase Coordination of Gamma Oscillations in CA1

Theta-frequency gamma-amplitude coupling analysis was useful to reveal distinct gamma oscillations in CA1. We next sought to study how the theta rhythm organizes those gamma oscillations in time. We next examined gamma power variation as a function of theta phase of the LFP recorded from CA1 str. pyramidale and filtered between 5 and 12 Hz (0° and 180° refer to positive polarity peak and negative polarity trough, respectively). LFP and IC's power was calculated for each frequency step from 30 to 300 Hz by complex wavelet transform. Z-scored gamma power for each frequency was plotted for each bin of theta phase to construct 2D comodulograms. Two theta cycles are represented for clarity.

For the LFP recorded from CA1 pyramidal layer we found three gamma sub-bands with distinct theta-phase distribution (Fig. 4.3a, first panel): gamma_S (indicated by one arrowhead) at the descending theta phase, gamma_M (two arrowheads) at the theta peak, gamma_F (three arrowheads) at the theta through. Theta-phase gamma-amplitude analysis applied to the ICs confirms this phase separation and clarifies the layer origin of each gamma component. Gamma_F power (>100 Hz)

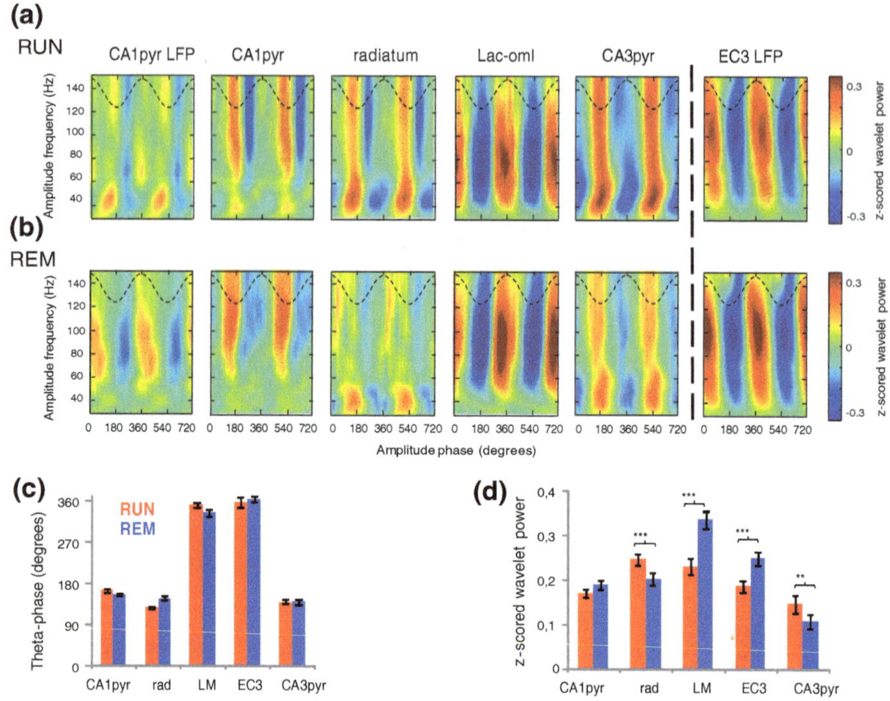

Fig. 4.3 **a** Gamma amplitude-theta phase modulation plots of LFP in CA1 pyramidal layer (*leftmost panel*) and CA1 LFP ICs during RUN. The gamma$_S$ (*single arrowhead*), gamma$_M$ (*double arrowheads*), and gamma$_F$ (*triple arrowhead*) dominated the descending phase, peak, and trough of the CA1 pyramidal layer theta waves, respectively. Dashed black line, reference theta phase of the LFP recorded in CA1 pyramidal layer. **b** Same as in (**a**) during REM. **c** Group data (six animals for CA1 and CA3, four rats for EC3) for preferred theta phase of each layer's theta-modulated gamma band (30–60, 60–110, and 100–250 Hz for *rad*, *lm*, and *CA1pyr*, respectively). **d** Z-scored theta-modulated gamma power across animals. (**/***$p < 0.01/0.001$, respectively; t test). Modified from [43]

dominated in the *CA1pyr* IC and was maximal near the trough of the theta cycle (174.8 ± 3.3°; Fig. 4.3a, c), coincident with the highest probability of spikes of the CA1 pyramidal cells and interneurons at this phase [13, 18, 35]. Gamma$_S$ (30–60 Hz) was most prominent in the *rad* IC, occurring predominantly on the descending phase of theta (128.3 ± 2.0°; Fig. 4.3a, c), which coincides with the preferred phase of gamma oscillations in the CA3pyr LFP (138.9 ± 4.5°; Fig. 4.3a, c), and most CA3 spiking [19, 35]. The gamma$_M$ (60–120 Hz) that dominated the *lm* IC was phase-locked to the peak of the reference theta waves (348.8 ± 5.3°; Fig. 4.3a, c), coincident with maximal EC3 gamma LFPs (355.8 ± 14.8°; Fig. 4.3a, c), and pyramidal cell firing in the entorhinal cortex L3 [35].

We also compared the theta phase distribution of gamma power during REM sleep. Although in the raw LFP was not evident the presence of the three gamma bands (Fig. 4.3b, first panel), all the ICs display very similar phase and frequency power distribution as during RUN. The theta phase and layer distributions of slow and mid-gamma activity was qualitatively similar to RUN (non-significant differences in theta phases of maximal gamma power, $p > 0.05$, t-test; Fig. 4.3b, c), but theta-modulated gamma power decreased in the *rad* and *CA3pyr* ICs ($p < 0.001$, t-test) while it increased in the *lm* IC and EC3 LFP ($p < 0.0001$, t-test) during REM (Fig. 4.3b, d).

As a next step, we further looked at the modulation of principal cell firing by the theta phase. For this purpose, we calculated the preferred theta phase for each excitatory neuron during RUN and REM in CA3, EC3 and CA1 (see Methods). The preferred firing phase of most CA3 pyramidal cells during both RUN and REM is at the descending phase of the theta cycle (Fig. 4.4, first panel), coinciding with the preferred phase of the radiatum $gamma_S$. In the case of the EC3 pyramidal cells the preferred firing phase in both states is around the peak of the theta cycle (Fig. 4.4, second panel), coinciding with the preferred phase of the lac-mol $gamma_M$. During RUN CA1 pyramidal cells tend to fire at the theta through or early ascending phase (Fig. 4.4, second panel), however during REM some cells shift

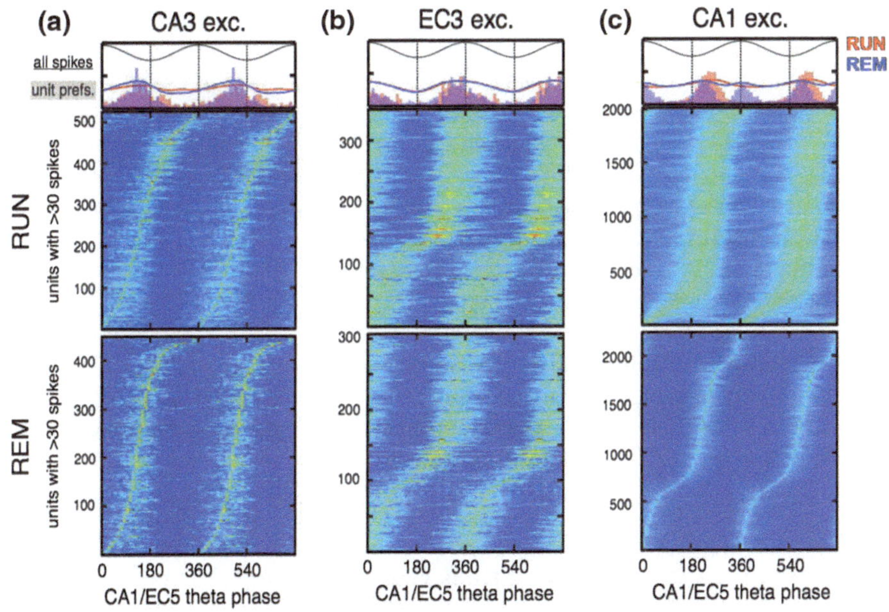

Fig. 4.4 Theta-phase modulation of pyramidal cells in CA3 (**a**), entorhinal cortex layer 3 (**b**) and CA1 (**c**) during RUN and REM. Units were sorted according to their z-scored theta-phase firing probability (*raster plots*). Histograms on the top represent the summed probability distribution of preferred phases and curves the summed probability distributions of firing rates (*red* for RUN and *blue* for REM) for all the units that were significantly theta-modulated (Rayleigh test $p < 0.01$). Black curve indicate reference theta phase in the CA1 pyramidal layer or entorhinal cortex layer 5. Note that the preferred phase for CA3 and EC3 pyramidal cells is similar to that obtained for rad and CA3pyr $gamma_S$ and lm and EC3 LFP $gamma_M$ (compare with Fig. 4.3a, b)

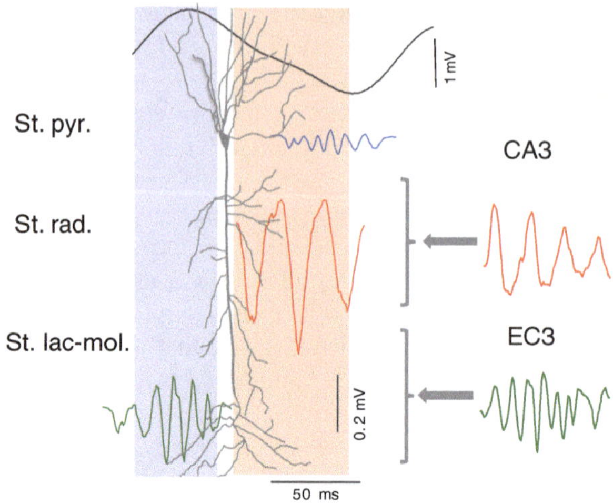

Fig. 4.5 Diagram summarizing the average ordering of the maximal phases for the gamma sub-bands, afferent input, and CA1 spike output over the theta cycle. Reproduced from [43]

their preferred phase toward the peak, reflecting an increased drive by EC3 input during this state [36].

The diagram in Fig. 4.5 summarizes the above results and proposed schematics for CA1 network function. At the theta peak, projection neurons in entorhinal cortex L3 increase their firing eliciting gamma$_M$ oscillations locally and in the CA1 stratum lacunosum-moleculare. After that, during the descending theta phase, CA3 pyramidal cells fire evoking gamma$_S$ oscillations also locally and in the CA1 stratum radiatum. The same CA3 axons that innervate pyramidal cell dendrites in the str. radiatum make abundant synapses onto CA1 interneurons. This feedforward inhibition may be responsible for the gamma$_F$ oscillations at the trough of the theta cycle. The firing probability of CA1 pyramidal cells is maximal around and after the theta through so it may be enhanced by the integration of coincident inputs from EC3 and CA3 and their exact timing controlled by the fast perisomatic inhibition, however this hypothesis still need to be demonstrated.

4.4 Variation of Theta-Coupled Gamma Oscillations Along the CA1 Transversal Axis

Along its transversal axis the CA1 region can be divided into a proximal region (closer to the CA2/CA3 border), an intermediate and a distal region (closer to the subiculum). There are important anatomical differences among them, as different entorhinal regions innervate the proximal and distal poles [45, 48]. It has also been suggested a functional specialization along the transverse axis. Place cells located in proximal sites are better and more spatially informative while those located in more distal sites have multiple place fields and less spatial coherence [28]. Less spatially selective cells in distal CA1 seemto be on the contrary more sensitive to the presence of objects [7] and odor cues [30]. In addition, it has been suggested that somatosensory information may also be processed differentially along the CA1 transverse axis [3]. This functional differences may be attributed to the innervation of proximal CA1 exclusively by MEC axons and distal CA1 by LEC inputs [48].

We studied the variation in theta-gamma dynamics along the CA1 transverse axis to check if they can offer a potential mechanism for the observed functional differences. For this purpose, we classified all the recording shanks from 6 animals as belonging to the proximal, intermediate or distal regions of CA1. Then we applied ICA to the LFPs recorded by each shank and performed theta-frequency gamma-amplitude analysis for the three main CA1 ICs, both during REM and RUN states.

Radiatum gamma$_S$ in proximal sites showed a stronger theta-gamma coupling than at distal sites, whereas the opposite relationship was observed for the gamma$_M$ band in the str. lacunosum-moleculare ($p < 0.001$ and $p <$, ANOVA tests; n = 6 rats; Fig. 4.6a). During REM sleep, *radiatum* gamma$_S$ power decreased in all sites while *lm* gamma$_M$ increased in all of them. However their spatial trends remained constant: gamma$_S$ power gradually decreasing from proximal to distal and gamma$_M$ increasing in the same direction (Fig. 4.6b). Conversely, the proximodistal distribution of gamma$_F$ in the pyramidal layer changes as a function of the brain state. During RUN the theta-gamma coupling of the pyramidal layer gamma$_F$ is stronger in proximal sites while during REM it shifted toward the opposite CA1 border ($p < 0.001$ and $p <$, ANOVA tests; n = 6 rats; Fig. 4.6a, b), coinciding with the trend of the predominant oscillation in each case, str. radiatum gamma$_S$ during RUN and str. lacunosum-moleculare gamma$_M$ during REM.

Changes in *CA1pyr* theta-gamma$_F$ coupling along the proximodistal axis and between RUN and REM mostly reflected changes in its theta phase distribution. A bimodal phase distribution, exhibiting increased power at both the trough and peak, emerged on the distal end during RUN and at all sites during REM (Fig. 4.7a). The altered phase distributions were limited to the *CA1pyr* ICs; quantifying the phase distribution with the 'center of mass' of gamma power showed significant variation in *CA1pyr* from proximal to distal ends during both RUN and REM ($p < 0.001$, ANOVA tests; Fig. 4.7b), but not in the other components ($p > 0.05$, ANOVA tests; Fig. 4.7b).

Regarding the observed variations in the strength of theta-gamma coupling and theta-phase distribution along the proximodistal axis, no significant change in the mean frequency of theta modulated gamma activity was observed for any of the three ICs in both states (Fig. 4.7c).

The differences found in the relative strength of the str. radiatum gamma$_S$, elicited by the CA3 input to the CA1, and the str. lacunosum-moleculare gamma$_M$, likely elicited by the EC3 input to CA1 pointed to a heterogeneous contribution of

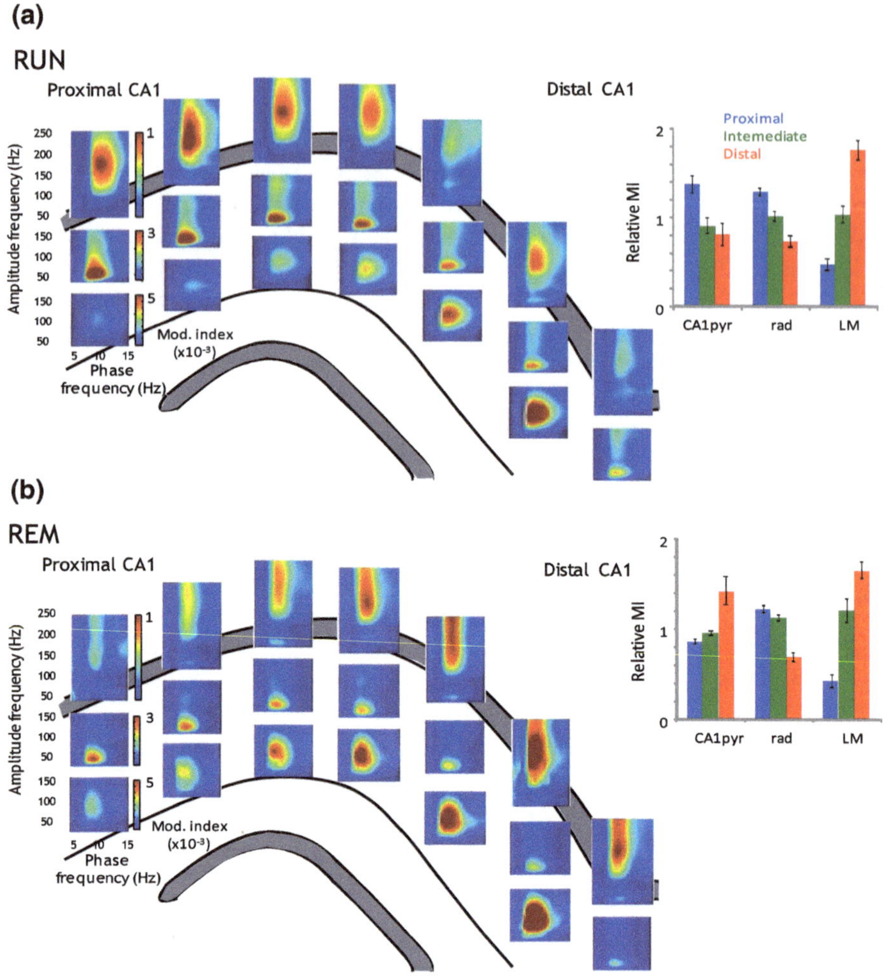

Fig. 4.6 **a** Two-dimensional distribution of the theta coupled gamma oscillations during RUN. Each set of three panels was constructed from the gamma-amplitude theta-phase comodulograms coupling of the ICs on each recording shank. The *rad* IC's gamma$_S$ became less strongly theta-modulated along the transversal (proximo-distal) axis, whereas lm gamma$_M$ increased its theta-coupled gamma power. CA1pyr gamma$_F$ largely followed the *rad* IC gradient. The bar plot on the right shows group data of MIs, normalized by the average across all shanks (six rats). Each comparison showed significant variation from proximal to distal sites ($p < 0.001$; ANOVA). **b** Similar display during REM. Note the opposite trend of CA1pyr theta-gamma$_F$ coupling along the proximo-distal axis compared to RUN. Reproduced from [43]

both inputs along the CA1 axis. CA3 inputs appear to be predominant in proximal sites and during RUN while EC3 inputs dominate in distal sites and during REM. Pyramidal layer gamma$_F$ varies along the axis according to the brain state, reflecting its variable modulation by both CA3 and EC3 inputs.

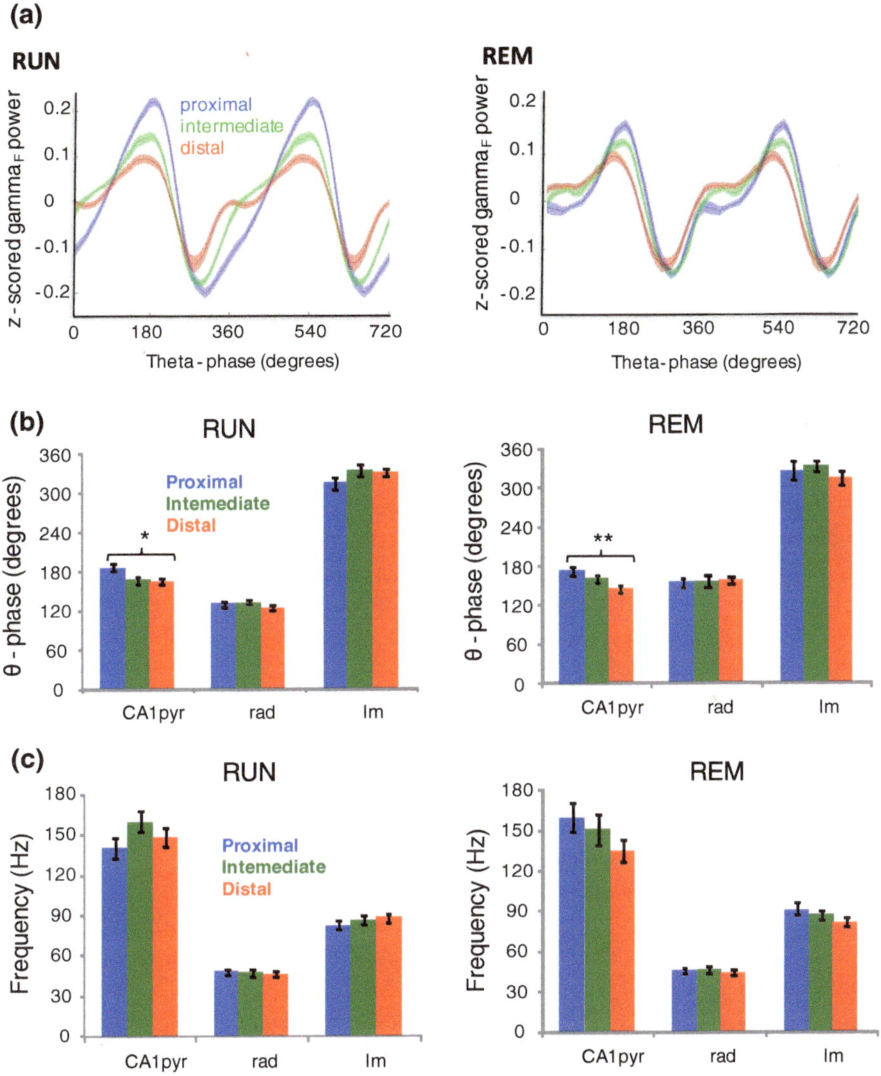

Fig. 4.7 **a** Theta phase distribution of CA1pyr gamma$_F$ power in RUN and REM at three segments along the transversal (proximo-distal) axis of CA1. Note increased bimodality of the gamma$_F$ power distribution (*arrowheads*) toward the distal end compared to the CA3 end ($p < 0.001$ for mean phases both in RUN and REM, ANOVA tests, six rats) and stronger overall bimodality during REM. **b** The 'centers of mass' of theta-phase (power-weighted mean phase) of *rad* gamma$_S$ and *lm* gamma$_M$ did not show significant changes along the proximo-distal axis in both RUN and REM ($p > 0.05$, t-test; 6 rats). However the *CA1pyr* gamma$_F$ shifts its preferred theta-phase from the theta trough at proximal CA1 to closer to the theta peak at distal sites in both RUN and REM (*/**$p < 0.05/0.01$, ANOVA tests; 6 rats). **c** Mean frequency of the three CA1 ICs (*CA1pyr*, *rad*, and *lm*) did not significantly change along the proximo-distal axis of the hippocampus in both RUN and REM states ($p > 0.05$, t-test; 6 rats). Modified from [43]

4.5 Theta-Gamma Cross-Frequency Coupling During a Memory Task

We have focused here on characterizing theta-gamma dynamics in the entorhino-hippocampal circuits and found different gamma oscillations organized by the theta rhythm. I also found that theta-gamma dynamics in the hippocampus displays strong state-dependent modulation during sleep and navigation; however they have also been implied in cognitive functions, as memory and learning [11, 34]. Thus I sought to test if the above characterized gamma oscillations have specific modulation during a hippocampus-dependent memory task. For this purpose I choose the delayed-alternation T-maze task [1, 38, 40]. In this task rats learn to run from a starting area through a central arm and then turn left or right, collect a water reward a return to the starting area (Fig. 4.8a). To increase their motivation for doing the task, rats are water deprived and can only drink during the task. After one of such trials they are forced to wait for 5–10 s and then start over again, but to get the reward animals have to turn to the opposite direction as in the previous trial. After a few days of training animals learn the task reaching a performance above 80 % (an error is when the animal chose to turn to the same direction as the previous trial). It has been shown that rodents with impaired hippocampal function cannot successfully perform this task [1]. In the central arm the animal is recalling the previous memory and uses this information to make a correct choice in the juncture of the maze. In the lateral arms the animal is encoding the current direction and has to keep this information in the working memory during the delay period [38, 49].

We compared the spectral activity of each of the three main CA1 LFPs ICs (*CA1pyr*, *radiatum* and *lac-mol*) while the rats were running in the central arm versus while they were running in the side arms. Only theta periods were included in the analysis and was checked that the velocity of the animal was not significantly different ($p > 0.5$ t-test) for the selected periods in each arm. Both theta-gamma coupling (Fig. 4.8b) and gamma power (Fig. 4.8d) were significantly stronger for each CA1 IC during center arm running, compared to the side arm with the largest changes present in CA1 str. radiatum ($p < 0.001$, t-test). These behavior-related changes were specific to the gamma sub-bands that dominate the respective ICs during theta: gamma_F (120–180 Hz) for CA1pyr, gamma_S (30–60 Hz) for radiatum and gamma_M (60–100 Hz) for lac-mol (Fig. 4.1c). These results indicate that theta-gamma coupling in the hippocampus is selectively enhanced during memory recall.

Theta-gamma analysis indicates that both CA3 and EC3 inputs are recruited during the recall phase of the task, although the CA3 input seems to be preferentially enhanced, as reflected by the largest increase in radiatum gamma_S. To further investigate if there is a predominance of CA3 input over the EC3 we checked the firing of the cells in the input and target regions in three additional animals with simultaneous hippocampus and entorhinal recordings, performing the same task in a slightly different maze (Fig. 4.9a). We aggregated spikes from each neuron type

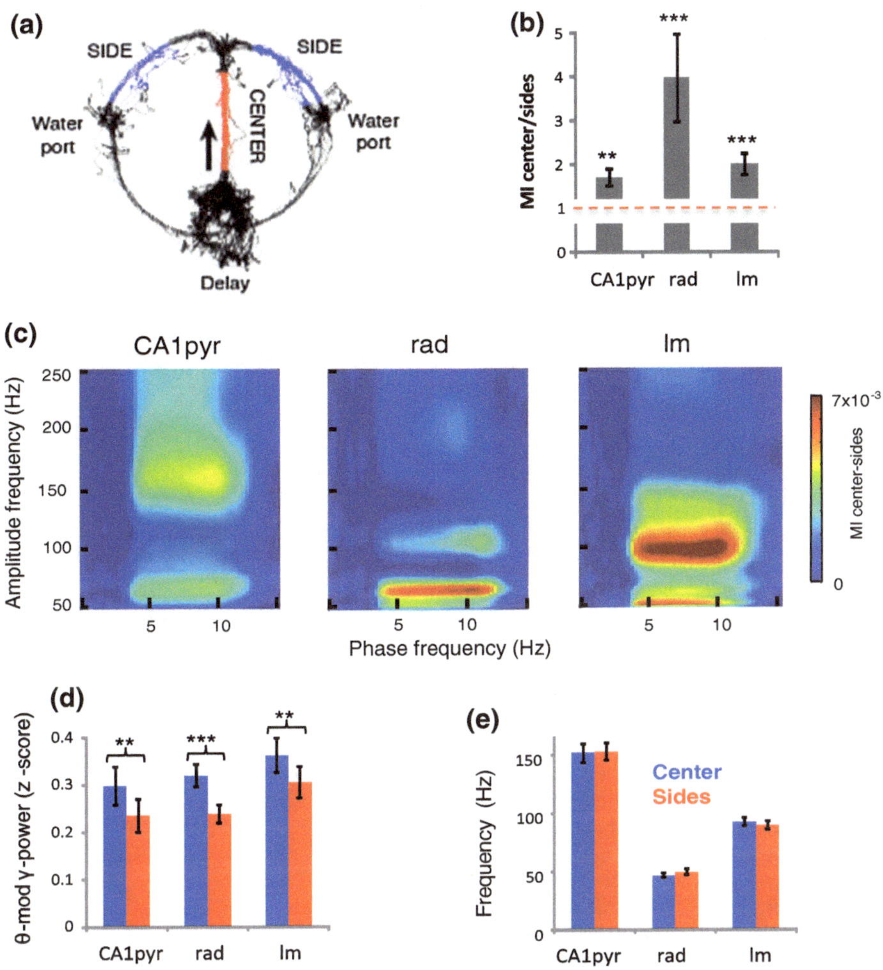

Fig. 4.8 **a** Example running trajectories during one session of a hippocampus-dependent delayed alternation T-maze task. Colors indicate sections choose to compare LFP and unit activity during center arm running (CENTER, *red*) to running in side arms (SIDE, *blue*). **b** The theta-gamma modulation indices (MI) for CA1 ICs were significantly greater during CENTER running compared to SIDE (t-test, n = 3 rats). **c** Differences in the theta modulation of the gamma power of each IC showed that the strongest changes were at the gamma frequencies characteristic of each afferent pathway (compare Fig. 4.1a). **d** Z-score-normalized power of each CA1 IC at the frequencies of strongest theta modulation were significantly greater during CENTER (*red*) running compared to SIDE (*blue*) (t-test; 3 rats). Changes were stronger for *rad* gamma$_S$. **E** Mean frequency of the three ICs did not significantly change between CENTER and SIDE running ($p > 0.05$; 3 rats). Modified from [43]

within each session to compare overall firing rates, because place-specific firing of hippocampal pyramidal cells would confound single unit comparisons of firing rates between arms [39]. CA3 pyramidal cell and CA1 interneuron firing rates were

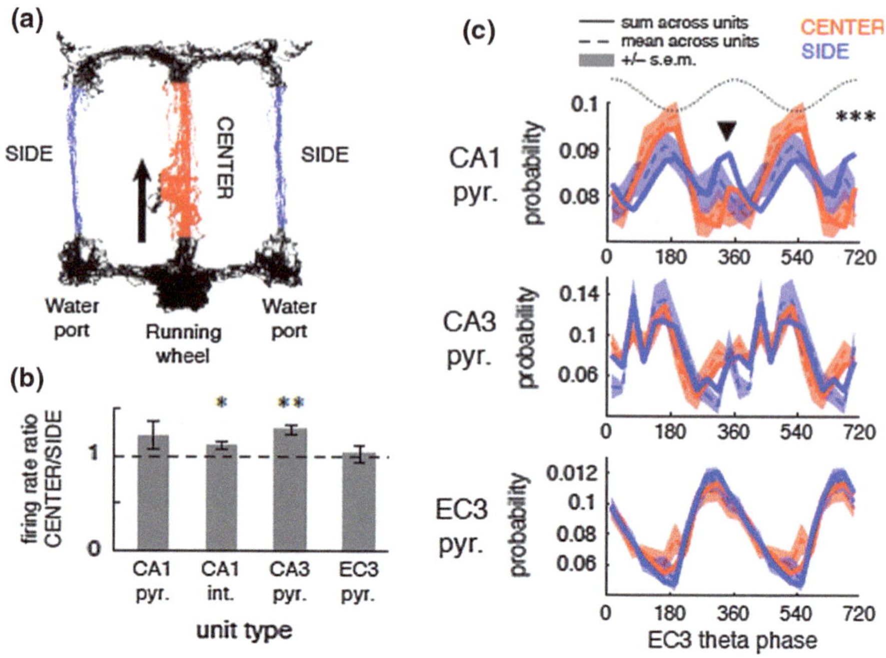

Fig. 4.9 **a** Example trajectories during one session of a similar delayed alternation task with wheel running during the delay period [40], which was performed by the animals used in the unit analyses. **b** Population firing rate ratios within individual sessions revealed that the CA3 pyramidal cells were significantly more active during CENTER running compared to SIDE (t-test). CA1 interneurons also fired at significantly greater rates in CENTER. Across sessions, the measured CA1 and EC3 pyramidal cell population rates were not consistently different in CENTER versus SIDE running ($p > 0.05$). **c** Theta phase (EC3 phase reference) distribution of spiking in CA1 (*top*), CA3 (*middle*), and EC3 (*bottom*) pyramidal cell populations during CENTER to SIDE running. *Lines* show the distribution when spike phases were summed across all clustered pyramidal cells, *dashed lines* (*shading*) show mean (±SEM) across single units firing > 30 spikes during CENTER and SIDE epochs. Significance thresholds: */**/***$p < 0.05/0.01/0.001$. Reproduced from [43]

significantly greater during center arm running compared to side arms ($p < 0.01$ for CA3 pyramidal cells, $p < 0.05$ for CA1 interneurons, t- test on center/side ratios; Fig. 4.9b). Although CA1 pyramidal cell firing rates were not significantly changed in the side arms compared to the center ($p > 0.5$ t-test, Fig. 4.9b) they showed a significantly greater probability of firing at the theta peak, the preferred phase of EC3 input (Fig. 4.9c, first panel). CA3 and EC3 pyramidal cells maintain their theta-phase preference constant in the center and side arms, the descending theta phase for the formers and the theta peak for the latter (Fig. 4.9c, second and third panels). Together with the increased firing of CA3 pyramidal cells in the central arm, this last result also pointed to an enhanced control of CA1 dynamics by CA3 input during memory recall.

Overall, these experiments demonstrate that the changing balance of CA3 and EC3 inputs during different phases of a hippocampus-dependent memory task can affect the theta-organization of both gamma activity and spiking in CA1.

References

1. Ainge JA, van der Meer MAA, Langston RF, Wood ER (2007) Exploring the role of context-dependent hippocampal activity in spatial alternation behavior. Hippocampus 17:988–1002
2. Amaral DG, Witter MP (1989) The three-dimensional organization of the hippocampal formation: a review of anatomical data. Neuroscience 31:571–591
3. Bellistri E, Aguilar J, Brotons-Mas JR, Foffani G, de la Prida LM (2013) Basic properties of somatosensory-evoked responses in the dorsal hippocampus of the rat. J Physiol 591: 2667–2686
4. Belluscio MA, Mizuseki K, Schmidt R, Kempter R, Buzsáki G (2012) Cross-frequency phase-phase coupling between θ and γ oscillations in the hippocampus. J Neurosci 32:423–435
5. Benito N, Fernandez-Ruiz A, Makarov VA, Makarova J, Korovaichuk A, Herreras O (2014) Spatial modules of coherent activity in pathway-specific LFPs in the hippocampus reflect topology and different modes of presynaptic synchronization. Cereb Cortex 24:1738–1752
6. Bragin A, Jandó G, Nádasdy Z, Hetke J, Wise K, Buzsáki G (1995) Gamma (40–100 Hz) oscillation in the hippocampus of the behaving rat. J Neurosci 15:47–60
7. Burke SN, Maurer AP, Nematollahi S, Uprety AR, Wallace JL, Barnes CA (2011) The influence of objects on place field expression and size in distal hippocampal CA1. Hippocampus 21:783–801
8. Buzsáki G (1984) Feed-forward inhibition in the hippocampal formation. Prog Neurobiol 22:131–153
9. Buzsáki G (2002) Theta oscillations in the hippocampus. Neuron 33:325–340
10. Buzsáki G, Wang X-J (2012) Mechanisms of gamma oscillations. Annu Rev Neurosci 35:203–225
11. Buzsáki G, Moser EI (2013) Memory, navigation and theta rhythm in the hippocampal-entorhinal system. Nature Publishing Group 16:130–138
12. Buzsáki G, Leung LS, Vanderwolf CH (1983) Cellular bases of hippocampal EEG in the behaving rat. Brain Res 287:139–171
13. Buzsáki G, Anastassiou CA, Koch C (2012) The origin of extracellular fields and currents—EEG, ECoG, LFP and spikes. Nat Rev Neurosci 13:407–420
14. Cabral HO, Vinck M, Fouquet C, Pennartz CMA, Rondi-Reig L, Battaglia FP (2014) Oscillatory dynamics and place field maps reflect hippocampal ensemble processing of sequence and place memory under NMDA receptor control. Neuron 81:402–415
15. Canolty RT, Edwards E, Dalal SS, Soltani M, Nagarajan SS, Kirsch HE, Berger MS, Barbaro NM, Knight RT (2006) High gamma power is phase-locked to theta oscillations in human neocortex. Science 313:1626–1628
16. Canolty RT, Knight RT (2010) The functional role of cross-frequency coupling. Trends in Cognitive Sciences 14:506–515
17. Colgin LL, Denninger T, Fyhn M, Hafting T, Bonnevie T, Jensen O, Moser M-B, Moser EI (2009) Frequency of gamma oscillations routes flow of information in the hippocampus. Nature 462:353–357
18. Csicsvari J, Hirase H, Czurkó A, Mamiya A, Buzsáki G (1999) Oscillatory coupling of hippocampal pyramidal cells and interneurons in the behaving rat. J Neurosci 19:274–287
19. Csicsvari J, Jamieson B, Wise KD, Buzsáki G (2003) Mechanisms of gamma oscillations in the hippocampus of the behaving rat. Neuron 37:311–322

20. Diba K, Buzsáki G (2008) Hippocampal network dynamics constrain the time lag between pyramidal cells across modified environments. J Neurosci 28:13448–13456
21. Fell J, Axmacher N (2011) The role of phase synchronization in memory processes. Nat Rev Neurosci 12:105–118
22. Fernández-Ruiz A, Herreras O (2013) Identifying the synaptic origin of ongoing neuronal oscillations through spatial discrimination of electric fields. Front Comput Neurosci 7:5
23. Fernández-Ruiz A, Schomburg EW (2013) The rules of entrainment: are CA1 gamma oscillations externally imposed or locally governed? J Neurosci 33:19045–19047
24. Fernández-Ruiz A, Makarov VA, Benito N, Herreras O (2012a) Schaffer-specific local field potentials reflect discrete excitatory events at gamma frequency that may fire postsynaptic hippocampal CA1 units. J Neurosci 32:5165–5176
25. Fernández-Ruiz A, Makarov VA, Herreras O (2012) Sustained increase of spontaneous input and spike transfer in the CA3-CA1 pathway following long-term potentiation in vivo. Front Neural Circ 6:71
26. Fernández-Ruiz A, Muñoz S, Sancho M, Makarova J, Makarov VA, Herreras O (2013) Cytoarchitectonic and dynamic origins of giant positive local field potentials in the dentate gyrus. J Neurosci 33:15518–15532
27. Glickfeld LL, Roberts JD, Somogyi P, Scanziani M (2009) Interneurons hyperpolarize pyramidal cells along their entire somatodendritic axis. Nat Neurosci 12:21–23
28. Henriksen EJ, Colgin LL, Barnes CA, Witter MP, Moser M-B, Moser EI (2010) Spatial representation along the proximodistal axis of CA1. Neuron 68:127–137
29. Ibarz JM, Foffani G, Cid E, Inostroza M, de la Prida LM (2010) Emergent dynamics of fast ripples in the epileptic hippocampus. J Neurosci 30:16249–16261
30. Igarashi KM, Lu L, Colgin LL, Moser M-B, Moser EI (2014) Coordination of entorhinal-hippocampal ensemble activity during associative learning. Nature 510:143–147
31. Lasztóczi B, Klausberger T (2014) Layer-specific GABAergic control of distinct gamma oscillations in the CA1 hippocampus. Neuron 81:1126–1139
32. Leung LS (1998) Generation of theta and gamma rhythms in the hippocampus. Neuroscience and biobehavioral reviews 22:275–290
33. Lisman JE, Idiart MA (1995) Storage of 7 +/− 2 short-term memories in oscillatory subcycles. Science 267:1512–1515
34. Lisman JE, Jensen O (2013) The theta-gamma neural code. Neuron 77:1002–1016
35. Mizuseki K, Sirota A, Pastalkova E, Buzsáki G (2009) Theta oscillations provide temporal windows for local circuit computation in the entorhinal-hippocampal loop. Neuron 64:267–280
36. Mizuseki K, Diba K, Pastalkova E, Buzsáki G (2011) Hippocampal CA1 pyramidal cells form functionally distinct sublayers. Nat Neurosci 14:1174–1181
37. Mizuseki K, Royer S, Diba K, Buzsáki G (2012) Activity dynamics and behavioral correlates of CA3 and CA1 hippocampal pyramidal neurons. Hippocampus 22:1659–1680
38. Montgomery SM, Buzsàki G (2007) Gamma oscillations dynamically couple hippocampal CA3 and CA1 regions during memory task performance. Proc Natl Acad Sci USA 104:14495–14500
39. O'Keefe J, Nadel L (1978) The Hippocampus as a Cognitive Map. Clarendon Press
40. Pastalkova E, Itskov V, Amarasingham A, Buzsáki G (2008) Internally generated cell assembly sequences in the rat hippocampus. Science 321:1322–1327
41. Scheffer-Teixeira R, Belchior H, Leão RN, Ribeiro S, Tort ABL (2013) On high-frequency field oscillations (> 100 Hz) and the spectral leakage of spiking activity. J Neurosci 33:1535–1539
42. Schomburg EW, Anastassiou CA, Buzsáki G, Koch C (2012) The spiking component of oscillatory extracellular potentials in the rat hippocampus. J Neurosci 32:11798–11811
43. Schomburg EW, Fernández-Ruiz A, Berényi A, Mizuseki K, Anastassiou CA, Koch C, Buzsáki G (2014) Theta phase segregation of input-specific gamma patterns in entorhinal-hippocampal networks. Neuron 84:470–485
44. Schroeder CE, Lakatos P (2009) Low-frequency neuronal oscillations as instruments of sensory selection. TINS 32:9–18
45. Steward O (1976) Topographic organization of the projections from the entorhinal area to the hippocampal formation of the rat. J Comp Neurol 167:285–314

46. Tort ABL, Komorowski R, Eichenbaum H, Kopell N (2010) Measuring phase-amplitude coupling between neuronal oscillations of different frequencies. J Neurophysiol 2:1195–1210
47. Vanderwolf CH (1969) Hippocampal electrical activity and voluntary movement in the rat. Electroencephalogr Clin Neurophysiol 26:407–418
48. Witter MP, Groenewegen HJ, Lopes da Silva FH, Lohman AH (1989) Functional organization of the extrinsic and intrinsic circuitry of the parahippocampal region. Prog Neurobiol 33: 161–253
49. Wood ER, Dudchenko PA, Robitsek RJ, Eichenbaum H (2000) Hippocampal neurons encode information about different types of memory episodes occurring in the same location. Neuron 27:623–633
50. Zemankovics R, Veres JM, Oren I, Hájos N (2013) Feedforward inhibition underlies the propagation of cholinergically induced gamma oscillations from hippocampal CA3 to CA1. J Neurosci 33:12337–12351

Chapter 5
Discussion

5.1 Methodological Challenges in the Study of Brain Oscillations

The study of brain oscillations is a rapidly expanding field. In recent years that it has been boosted by the development of new technologies to perform large-scale recordings and manipulations of brain activity in intact organisms. The increased volume and complexity of data typically obtained in a Systems Neuroscience experiment performed with recent methods poses an important challenge for their analysis and interpretation. From our point of view, the development of novel analytical approaches has been lagging behind the technical development in this field. For this reasons one of the main goals of the present work has been to explore new methodological approaches to the analysis of neural circuit function and in particular brain oscillations.

The most traditional approach to study brain oscillations is to filter in the recorded signals in narrow frequency bands to them decompose them in the spectral domain by means of Fourier, wavelet or similar transform. Spectral characteristics of the signal (power or phase in a certain frequency band) can be then related to behavior or other external variables. This simple approach has been very successful in a multitude of occasions. Some classical examples are identification of the different phases of sleep regarding the spectral profile of the EEG [17], the correlation of visual stimulus properties with visual cortex gamma oscillations [31] or the modulation of hippocampal theta frequency and power by movement speed [8]. However this approach is often too simplistic to capture the variability of complexity of neural oscillatory responses that are rarely narrow band restricted or stationary in time. Numerous recent reports point to the variability of oscillations during behaviour and suggest that the classical view of adcrisbing cognitive functions to particular frequency bands is no longer tenable [23, 27, 58, 61]. We found an example of this controversy in the role adscribed to neocortical gamma oscillations as a binding mechanism for different visual features into an

A. Fernández-Ruiz, *Extracellular Potentials in the Hippocampus*,
Springer Theses, DOI 10.1007/978-3-319-41039-5_5

unified perception [29, 64]. Other studies challenge this assumption arguing that the response to stimuli in the visual cortex largely varies in frequency at different points of the same area [57]. We are moving from a view in which strictly spectrally defined oscillations are responsibles for specific cognitive functions to recognize the intrinsic variability and nested organization of brain oscillations that can flexibly support multiple computational process. As an example that we can not talk about “gamma rhythm” in general, 40–50 Hz oscillations in the visual cortex [31, 64] or the hippocampus [5, 16, 24] may not share common generating mechanisms and definitely serve to different functions. These issues stress the importance of identifying the population/s contributing to a given field potential oscillation as a necessary step to infer on its physiological and computational meaning. In summary, to interpret spectral changes in population activity during behavior we need to know the cellular generators of observed oscillations as well as their organizing network mechanisms.

These problems are more easily addressed with invasive recording techniques, in particular those that allow a dense mapping of the structure of interest in the three spatial dimensions, such as large scale silicon-probe recordings. In the case of non-invasive recordings such as EEG or MEG the problem become much complicated. We understand the basic biophysical properties of electric activity propagation in the brain tissue but our current knowledge of how the synaptic activity scales to the population level is very far from being comprehensive. To bridge this gap we propose that, in the first place, is necessary to understand the cellular mechanisms of local oscillatory activity generation, by means of careful LFP and single unit recordings and analysis. As a second step we need to employ macroscopic models to understand how local activity is implemented at the whole structure level and how the cytoarchitectonic and functional characteristic of a particular cellular population influence the generation and propagation of currents on the tissue [26].

One of the most common approaches to address the inverse problem of the LFP, i.e. to determine the subcellular currents that elicit a particular voltage distribution, is the current source density (CSD) analysis [28]. Linear silicon probes are ideal to perform this analysis as they allow a dense mapping of a small region with a well-defined geometric arrangement. This approach has been very successful to unveil the underlying currents of some of the most common electrophysiological patterns such as evoked potentials [45], sharp-waves [73] or hippocampal theta rhythm [6]. However due to the fact that most LFP patterns are elicited by the mixed contribution of multiple sources in most of the cases CSD analysis alone is not conclusive to understand their underlying mechanisms [23].

5.2 A Novel Approach to the Study of Brain Oscillations

Amongst the approaches used to address the mixed contribution of inputs to macroscopic patterns, some sought the selective manipulation of parts of a network, such as the activation or the silencing of specific pathways or neuron types through electrical, optogenetic, or pharmacological intervention [33, 42, 70]. Other approaches pursued the disentanglement of LFPs into their original generators by applying statistical tools and algorithms [18, 22, 53, 40]. Blind source separation techniques, like the independent component analysis (ICA: [13, 15]), appear to be the best suited by their capacity to find stable groups of sensors picking up a signal whose origin is stationary in the space, a feature that can be assumed for electrical fields generated by synaptic currents. We have developed an implementation based on the ICA of depth profiles of LFP followed by CSD analysis of the ICs to separate the different synaptic pathways converging on hippocampal principal cells on the evidence that each produces field potentials of stable and distinct spatial distribution [23, 24]. Applying this method to two-dimensional LFP recordings in the hippocampus we were able separate their physiologically meaningful sources and match them with known anatomical pathways and its dendritic domains. Due to the parallel arrangement of principal cells bodies and dendrites in the hippocampus and the stratification of the synaptic inputs, the application of ICA and the interpretation of the resulting ICs are particularly simple. However it has some limitations like only those synaptic inputs with enough postsynaptic current, synchrony, and spatial clustering can be detected and separated by ICA; thus, very weak or sparse currents are not easily discernible [26].

Temporal and spectral characteristics of the wide-band ICs can be analyzed. We thus can detect arbitrary frequency restricted oscillatory patterns without necessary imposing a previous narrow-band filter to the LFP. Another advantage is that the use of pathway-specific ICs reduced the concern of a particular LFP pattern to be a mixture of several inputs, allowing a more accurate interpretation of oscillations in terms of their underlying mechanisms [63].

We have also employed in this work an innovative methodology to simulate LFPs at a mesoscopic scale. It consists in explicitly modelling the geometry of the tissue, simulating neurons as current sources whose weight and temporal activation was taken from experimental measures and solving the electromagnetic differential equations to calculate electric fields and potentials with FEM. Despite its simplicity, the model was able to reproduce with high accuracy the characteristics of experimentally recorded hippocampal LFPs. The experimental data necessary to correctly implement the model was only available by the use multielectrode recordings and the ability of ICA to render spatial loadings and temporal activations of LFP sources (that is, to solve the inverse problem of LFPs). Previous approaches used realistic connectivity and/or membrane electrogenesis to explore the intracellular factors and population dynamics that produce significant amount of transmembrane currents [46, 55, 62]. The advantage of the FEM approach is that explicit modeling of the population's architecture and the conducting volume allows investigating whether or not these currents build macroscopic LFPs. The FEM approach is widely used in MEG and scalp EEG [12, 60, 66], but to our knowledge this has not been used before in LFP studies.

The present work highlights the importance of tissue geometry in the spreading of field potentials. Thus, the question of the spatial reach of LFPs [39, 41, 69, 71] will not have a unique answer; it will depend on the particular architecture of the region where they are recorded as well as the synchrony of the presynaptic neurons. These factors cannot be properly investigated without realistic modelling of spatial factors, which turn to be essential to understand and interpret LFPs.

The conclusion drawn in this work could be potentially applicable to LFPs generated in other structures with similar curved geometrical characteristics. That is the case of cortical sulcus in humans and other mammals in which cortical layers bend to form giri and sulci. LFPs in the cortex are in general of lower amplitude than in the hippocampus due to the dispersion of cell bodies in comparison with the densely packed cellular layers in hippocampus, which leads to a partial cancellation of extracellular currents. Despite this difference, which will lead to important quantitative alterations in LFPs magnitude, it seems plausible that the same geometrical and dynamical factors that we have shown, determine the characteristics of DG LFPs will rule also for potentials inside cortical sulci. These considerations may have special importance for MEG, due to fact that the principal contributors to MEG signals are the neurons located with their main axis perpendicular to the cortical surface, which are mainly those situated in the sulci [47, 54].

5.3 Theta-Gamma Dynamics Reveals Network Computations During Behaviour

The results presented in the third section of the Results demonstrate that low frequency gamma oscillations link together the CA3-CA1 regions and that the strength of this relationship is strongest during memory recall, associated with increased theta coupled gamma_S power and a shift of preferred spike theta phase from the peak to near the trough in a fraction of CA1 pyramidal cells. A distinct and higher frequency gamma pattern (gamma_M) is dominant near the peak of the theta cycle, coincident with increased firing of EC3 pyramidal neurons and their excitation of the distal apical dendrites of CA1 pyramidal cells (current sink). Gamma coherence is high within the same dendritic layer but low across layers. During REM sleep CA3 spiking output and the associated gamma_S in CA1 str. radiatum are decreased, accompanied by an increased EC3 spiking and elevated gamma_M power in str. lacunosum-moleculare.

Our findings support previous observations that gamma_S oscillations occur on the descending phase of theta in the CA1 pyramidal layer, brought about by the gamma-timed spiking of CA3 pyramidal cells [3, 16, 24, 25, 43]. However, they are at variance with the postulated theta phase assignment of EC3-mediated gamma bursts by Colgin et al. [14]. In the latter study, the authors suggested that the CA3-driven gamma_S is *followed* by an EC input-driven faster gamma burst in the

65–140 Hz range at the trough of the theta cycle [14]. One potential source of the conflicting results is the use of single site recordings in the Colgin et al. study, which is not sufficient to decompose the spatial origin of the respective gamma generators [9, 21, 23, 43]. Because the main generators of LFPs are often postsynapic currents, a single oscillating population may generate coherent LFPs in distant areas without necessarily entraining the downstream population. Although the authors showed phase coupling between entorhinal unit firing and the lumped gamma_M and gamma_F band LFPs in the CA1 pyramidal layer, they did not report the theta phase of either spiking or gamma activity in EC3. In contrast to the framework put forward by Colgin et al. [14], we find that the peak firing of EC3 pyramidal cells and the associated current sink in the CA1 str. lacunosum-moleculare occurs, on average, at the peak of the theta cycle [4, 6, 10, 38, 50, 53], coincident with the gamma_M burst and, thus, *before* the dominant CA3-mediated gamma_S on the descending phase of theta.

The differential firing rate changes of CA3 and EC3 principal cell can account for the CA1 firing patterns and gamma oscillations, we observed both across brain states and in the T-maze task. These two major inputs can either compete or cooperate depending on the nature of the task [2, 20, 37, 44, 59]. The EC3 input may become more effective through theta-rhythmic frequency potentiation by suppressing feed-forward inhibition [11, 44, 72] or facilitating mechanisms for overcoming it [37, 59, 65]. Furthermore, during REM sleep, firing rates of CA3 pyramidal neurons decrease [48, 52], paralleled with the reduced power of gamma_S in CA1 str. radiatum and decreased spike-field CA3-CA1 coherence [63]. This reduced CA3 output coincides with an increased firing of EC3 neurons and elevated gamma_M power in the str. lacunosum-moleculare and the shifting of theta phase preference of a significant fraction of CA1 pyramidal neurons from the trough to the peak [49, 56]. Thus, a weakened CA3 output leads to a relatively stronger control of the EC3 input on the discharge of CA1 neurons, as was also observed in the side arms of the T-maze task, where demands on memory retrieval are diminished.

An increased gain control over CA1 neurons by the direct EC3 input can explain why after lesion of the CA3 input, place-related firing of CA1 pyramidal cells can persist [7]. The shift in gain control can also explain why spatial information encoded by CA1 pyramidal cells varies as a function of their position in the proximodistal axis [35] and why during recall the CA3 input shows increased control over CA1 activity patterns [51]. One potential substrate of the CA3-EC3 competition is the O-LM to bistratified interneuron inhibitory connection. Increased firing of O-LM interneurons near the theta trough [67] suppresses inputs to the distal dendrites, whereas itindirectly disinhibits dendritic segments in str. oriens and radiatum [44], thereby facilitating the effectiveness of the CA3 input to CA1 pyramidal cells.

During encoding of newly learned information, EC input is expected to boost the efficacy of the CA3-CA1 pathway [34, 59]. Indeed, in vitro experiments have demonstrated that pairing temporo-ammonic (EC3 to CA1) and Schaffer collateral excitation to CA1 can multiplicatively induce long-term potentiation and CA1 spiking via dendritic plateau potentials and heterosynaptic effects, depending on the relative frequency and timing of these inputs [2, 11, 19, 30, 32, 59, 65]; but see [1]. The strongest potentiation is observed when the distal dendrites are excited 10–30 ms prior to the arrival of the CA3 input, matching the roughly one quarter theta cycle delay between the occurrence of gamma_M and gamma_S in CA1. The heterosynaptic boosting effect of the entorhinal input can be selectively suppressed by perisomatic inhibition mediated by cholecystokinin (CCK)-expressing interneurons [2]. Our findings demonstrate that the timing of these spike level interactions can be monitored by the mesoscopic changes of the LFP theta-gamma oscillations. Additional experiments are required to disentangle the roles of the medial and lateral entorhinal cortical inputs [36] and the thalamic nucleus reuniens [68] in the expression of gamma in the str. lacunosum-moleculare.

Overall, our experiments demonstrate that layer-specific gamma oscillations in the hippocampus reliably identify the temporal dynamics of the afferent inputs and that temporal coordination in the entorhinal-hippocampal system is mainly supported by theta and low frequency gamma oscillations, but not by high gamma coherence [63].

5.4 Conclusions

We have implemented a novel method to the study of LFPs and applied it to hippocampal recordings from behaving rats. The first important conclusion we get from this work is that for a correct analysis and interpretation of LFPs it is necessary an appropriated spatial mapping of the structure of interest. Due to the propagation of electric potentials in the brain tissue, single point recordings, as traditionally performed with tetrodes or wire electrodes, are not able to discern the origin (local, propagated, mixed) of the recorded LFPs. On the contrary, high-density two-dimensional recordings, as those presented here, are optimal because they account for the spatial variations of LFPs in laminated structures such as the hippocampus or neocortex.

We employed ICA to decompose hippocampal LFPs recorded along the dorsoventral and transversal axes of the hippocampus into six main contributing sources (independent components, ICs). Those sources display distinct laminar distribution and their CSD maps reveal main currents restricted to particular sub-layers, allowing us to identify their synaptic nature comparing with known anatomical synaptic domains. The three main CA1 ICs were identified as the ipsilateral CA3 input to the str. radiatum dendrites (*rad* IC), the entorhinal cortex layer 3 input to the str. lacunosum-moleculare dendrites (*lm* IC) and perisomatic currents, likely inhibitory,in the CA1 pyramidal cell bodies (*CA1pyr* IC). In the DG they were identified as the lateral perforant path input to the distal GC dendrites (*LPP* IC), the medial perforant path input to the middle GC dendrites (*MPP* IC), and perisomatic currents in the GC somas, likely inhibitory, (*GCsom*). However,a more rigorous identification of those ICs as strictly pathway-specific or as a mixture of different inputs with shared dendritic domains would require additional tests, like selective optogenetic manipulations in well-defined cellular populations. This is especially relevant for the *CApyr* and *GCsom* components, which likely capture currents elicited by somatic action potentials in addition to postsynaptic currents.

Simulations performed with FEM in the tridimensional model of the hippocampus reproduce with high fidelity the spatial distribution of the different hippocampal current sources. The simultaneous activation of the six sources with time series obtained from experimental recordings allows us to reproduce the laminar characteristics of SWR and theta oscillations. These results confirm the accuracy of our solution for the inverse problem of hippocampal LFPs and that those six sources are enough to account for most of the variability of the recorded LFPs in the CA1 and DG regions. We have decided not to include here other results pertaining to the CA3 were another prominent IC can be isolated [4, 63].

The above results illustrate the usefulness of our novel method to analyze LFPs. It consists in first extract and identify the current sources contributing to their generation and then implement the obtained solution in a model that allows the forward simulation of LFPs. By comparing the original and simulated LFPs we can refine and better interpret our original solution.

In the second section of the Results, we employed the acquired knowledge about the sources of hippocampal LFPs to address the study of theta-gamma interactions in the hippocampal-entorhinal circuit during behaviour. We found that the three main LFP sources characterized in the first section of the Results display different spectral profiles in the gamma band and theta-phase preference. Slow gamma (gamma_S, 30–60 Hz) was elicited by the CA3 input to the CA1 and peaked on the descending theta phase, while mid-frequency gamma (gamma_M, 60–120 Hz) brought about by entorhinal layer 3 input dominated on the peak of the theta cycle. A third source of very fast oscillations (gamma_F, 100–180 Hz) peaked on the theta trough and was of local origin. The preferred phase of *rad* slow gamma was shared by similar frequency oscillations in the input region, CA3, and was also the theta-phase with higher probability of firing of CA3 pyramidal cells. Correspondingly, the preferred phase of *lm* gamma_M was shared by similar frequency gamma oscillations in the layer 3 of the medial entorhinal cortex and was also the theta phase with higher probability of discharge of EC3 pyramidal cells. Those theta-coupled gamma oscillations were present in both exploration and REM sleep albeit with different strength. Theta-coupled gamma oscillations not only vary as a function of behavioral state but also with the location along the CA1 transversal axis. CA3 gamma_S was stronger during running and dominated at proximal sites, while gamma_M was stronger during REM and dominated at distal CA1 sites. Perisomatic gamma_F oscillations varied in their proximodistal distribution and theta-phase preference as a function of brain state according to which gamma_S or gamma_M was the dominant pattern. This illustrates a competition of CA3 and entorhinal inputs to control CA1 local circuits operation.

We also checked theta-gamma dynamics in CA1 during memory guided navigation (a delayed-alternation T maze task known to relay on the hippocampus). We found that during the recall phase of the task (the central arm running) theta-gamma coupling increased for all the ICs but especially for the *rad* $gamma_S$. In addition, during this phase the firing rate of CA3 pyramidal cells was increased. On the contrary, at the encoding phase of the task (side arms running) CA3 firing rates were lower and a subset of CA1 pyramidal cells shift their theta phase preference to the theta peak, the phase of the EC3 input. Those results demonstrate that CA3 and EC3 inputs can compete or cooperate to control the firing of CA1 cells according to behavioral demands.

References

1. Ang CW, Carlson GC, Coulter DA (2005) Hippocampal CA1 circuitry dynamically gates direct cortical inputs preferentially at theta frequencies. J Neurosci 25:9567–9580
2. Basu J, Srinivas KV, Cheung SK, Taniguchi H, Huang ZJ, Siegelbaum SA (2013) A cortico-hippocampal learning rule shapes inhibitory microcircuit activity to enhance hippocampal information flow. Neuron 79:1208–1221
3. Belluscio MA, Mizuseki K, Schmidt R, Kempter R, Buzsáki G (2012) Cross-frequency phase-phase coupling between θ and γ oscillations in the hippocampus. J Neurosci 32: 423–435
4. Benito N, Fernandez-Ruiz A, Makarov VA, Makarova J, Korovaichuk A, Herreras O (2014) Spatial modules of coherent activity in pathway-specific LFPs in the hippocampus reflect topology and different modes of presynaptic synchronization. Cereb Cortex 24:1738–1752
5. Bragin A, Jandó G, Nádasdy Z, Hetke J, Wise K, Buzsáki G (1995) Gamma (40–100 Hz) oscillation in the hippocampus of the behaving rat. J Neurosci 15:47–60
6. Brankack J, Stewart M, Fox SE (1993) Current source density analysis of the hippocampal theta rhythm: associated sustained potentials and candidate synaptic generators. Brain Res 615:310–327
7. Brun VH, Leutgeb S, Wu H-Q, Schwarcz R, Witter MP, Moser EI, Moser M-B (2008) Impaired spatial representation in CA1 after lesion of direct input fromentorhinalcortex. Neuron 57:290–302
8. Buzsáki G, Leung LW, Vanderwolf CH (1983) Cellular bases of hippocampal EEG in the behaving rat. Brain Res 287:139–171
9. Buzsáki G, Anastassiou CA, Koch C (2012) The origin of extracellular fields and currents—EEG, ECoG, LFP and spikes. Nat Rev Neurosci 13:407–420
10. Buzsáki G, Czopf J, Kondakor I, Kellenyi L (1986) Laminar distribution of hippocampal rhythmic slow activity (RSA) in the behaving rat: current source density analysis, effects of urethane and atropine. Brain Res 365:125–137
11. Buzsáki G, Penttonen M, Bragin A, Nadasdy Z, Chrobak JJ (1995) Possible physiological role of the perforant path-CA1 projection. Hippocampus 5:141–146
12. Chen M and Mogul DJ (2009) A structurally detailed finite element human head model for simulation of transcranial magnetic stimulation. J Neurosci Meth 179:11–120
13. Choi S, Cichocki A, Park HM, Lee SY (2005) Blind source separation and independent component analysis: a review. Neural Inf Process Lett Rev. 6:1–57
14. Colgin LL, Denninger T, Fyhn M, Hafting T, Bonnevie T, Jensen O, Moser M-B, Moser EI (2009) Frequency of gamma oscillations routes flow of information in the hippocampus. Nature 462:353–357
15. Comon P (1994) Independent component analysis. A new concept? Sig Process 138:287–314

16. Csicsvari J, Jamieson B, Wise KD, Buzsáki G (2003) Mechanisms of gamma oscillations in the hippocampus of the behaving rat. Neuron 37:311–322
17. Dement W, Kleitman N (1957) Cyclic variations in EEG during sleep and their relation to eye movements, body motility and dreaming. Electroenceph. Clin Neurophysiol 9:673–690
18. Di S, Baumgartner C, Barth DS (1990) Laminar analysis of extracellular field potentials in rat vibrissa/barrel cortex. J Neurophysiol 63:832–840
19. Dudman JT, Tsay D, Siegelbaum SA (2007) A role for synaptic inputs at distal dendrites: instructive signals for hippocampal long-term plasticity. Neuron 56:866–879
20. Dvorak-Carbone H, Schuman EM (1999) Long-term depression of temporoammonic-CA1 entorhinal cortex: redefinition of the role of the trisynaptic pathway. Proc Natl Acad Sci USA 87:5832–5836
21. Einevoll GT, Kayser C, Logothetis NK, Panzeri S (2013) Modelling and analysis of local field potentials for studying the function of cortical circuits. Nat Rev Neurosci 14:770–785
22. Einevoll GT, Pettersen KH, Devor A, Ulbert I, Halgren E, Dale AM (2007) Laminar population analysis: estimating firing rates and evoked synaptic activity from multielectrode recordings in rat barrel cortex. J Neurophysiol 97:2174–2190
23. Fernández-Ruiz A, Herreras O (2013) Identifying the synaptic origin of ongoing neuronal oscillations through spatial discrimination of electric fields. Front Comput Neurosci 7:5
24. Fernández-Ruiz A, Makarov VA, Benito N, Herreras O (2012) Schaffer-specific local field potentials reflect discrete excitatory events at gamma frequency that may fire postsynaptic hippocampal CA1 units. J Neurosci 32:5165–5176
25. Fernández-Ruiz A, Makarov VA, Herreras O (2012) Sustained increase of spontaneous input and spike transfer in the CA3-CA1 pathway following long-term potentiation in vivo. Front Neural Circuits 6:71
26. Fernández-Ruiz A, Muñoz S, Sancho M, Makarova J, Makarov VA, Herreras O (2013) Cytoarchitectonic and dynamic origins of giant positive local field potentials in the dentate gyrus. J Neurosci 33:15518–15532
27. Florian G, Andrew C, Pfurtscheller G (1998) Do changes in coherence always reflect changes in functional coupling? Electroencephalogr Clin Neurophysiol 106:87–91
28. Freeman JA, Nicholson C (1975) Experimental optimization of current source-density technique for anuran cerebellum. J Neurophysiol 38:369–382
29. Fries P, Nikolic D, Singer W (2007) The gamma cycle. Trends Neurosci 30:309–316
30. Golding NL, Mickus TJ, Katz Y, Kath WL, Spruston N (2005) Factors mediating powerful voltage attenuation along CA1 pyramidal neuron dendrites. J Physiol 568:69–82
31. Gray CM, König P, Engel AK, Singer W (1989) Oscillatory responses in cat visual cortex exhibit inter-columnar synchronization which reflects global stimulus properties. Nature 338:334–337
32. Han EB, Heinemann SF (2013) Distal dendritic inputs control neuronal activity by heterosynaptic potentiation of proximal inputs. J Neurosci 33:1314–1325
33. Happel MF, Jeschke M, Ohl FW (2010) Spectral integration in primary auditory cortex attributable to temporally precise convergence of thalamocortical and intracortical input. J Neurosci 30:11114–11127
34. Hasselmo ME, Bodelón C, Wyble BP (2002) A proposed function for hippocampal theta rhythm: separate phases of encoding and retrieval enhance reversal of prior learning. Neural Comput 14:793–817
35. Henriksen EJ, Colgin LL, Barnes CA, Witter MP, Moser M-B, Moser EI (2010) Spatial representation along the proximodistal axis of CA1. Neuron 68:127–137
36. Igarashi KM, Lu L, Colgin LL, Moser M-B, Moser EI (2014) Coordination of entorhinal hippocampal ensemble activity during associative learning. Nature 510:143–147
37. Jarsky T, Roxin A, Kath WL, Spruston N (2005) Conditional dendritic spike propagation following distal synaptic activation of hippocampal CA1 pyramidal neurons. Nat Neurosci 8:1667–1676

38. Kamondi A, Acsády L, Wang XJ, Buzsáki G (1998) Theta oscillations in somata and dendrites of hippocampal pyramidal cells in vivo: activity-dependent phase-precession of action potentials. Hippocampus 8:244–261
39. Katzner S, Nauhaus I, Benucci A, Bonin V, Ringach DL, Carandini M (2009) Local origin of field potentials in visual cortex. Neuron 61:35–41
40. Kocsis B, Bragin A, Buzsáki G (1999) Interdependence of multiple theta generators in the hippocampus: a partial coherence analysis. J Neurosci 19:6200–6212
41. Kreiman G, Hung CP, Kraskov A, Quiroga RQ, Poggio T, DiCarlo JJ (2006) Object selectivity of local field potentials and spikes in the macaque inferior temporal cortex. Neuron 49:433–445
42. Kuki T, Ohshiro T, Ito S, Ji ZG, Fukazawa Y, Matsuzaka Y et al (2012) Frequency-dependent entrainment of neocortical slow oscillation to repeated optogenetic stimulation in the anesthetized rat. Neurosci Res 75:35–45. doi:10.1016/j.neures.2012.10.007. (Epub ahead of print)
43. Lasztóczi B, Klausberger T (2014) Layer-specific GABAergic control of distinct gamma oscillations in the CA1 hippocampus. Neuron 81:1126–1139
44. Leão RN, Mikulovic S, Leão KE, Munguba H, Gezelius H, Enjin A, Patra K, Eriksson A, Loew LM, Tort ABL, Kullander K (2012) OLM interneurons differentially modulate CA3 and entorhinal inputs to hippocampal CA1 neurons. Nat Neurosci 15:1524–1530
45. Leung LS (1979) Potentials evoked by alvear tract in hippocampal CA 1 region in rats. II. Spatial field analysis. J Neurophysiol 42:1571–1589
46. Lindén H, Tetzlaff T, Potjans TC, Pettersen KH, Gru S, Diesmann M, Einevoll G (2011) Modeling the spatial reach of the LFP. Neuron 859–872
47. Malmivuo J, Plonsey R (1995) Bioelectromagnetism: principles and applications of bioelectric and biomagnetic fields. Oxford University Press, New York
48. Mizuseki K, Buzsáki G (2013) Preconfigured, skewed distribution of firing rates in the hippocampus and entorhinal cortex. Cell Rep 4:1010–1021
49. Mizuseki K, Diba K, Pastalkova E, Buzsáki G (2011) Hippocampal CA1 pyramidal cells form functionally distinct sublayers. Nat Neurosci 14:1174–1181
50. Mizuseki K, Sirota A, Pastalkova E, Buzsáki G (2009) Theta oscillations provide temporal windows for local circuit computation in the entorhinal-hippocampal loop. Neuron 64: 267–280
51. Montgomery SM, Buzsàki G (2007) Gamma oscillations dynamically couple hippocampal CA3 and CA1 regions during memory task performance. Proc Natl Acad Sci USA 104: 14495–14500
52. Montgomery SM, Sirota A, Buzsáki G (2008) Theta and gamma coordination of hippocampal networks during waking and rapid eye movement sleep. J Neurosci 28:6731–6741
53. Montgomery SM, Betancur MI, Buzsáki G (2009) Behavior-dependent coordination of multiple theta dipoles in the hippocampus. J Neurosci 29:1381–1394
54. Nunez PL, Srinivasan R (2006) Electric Fields of the Brain: The Neurophysics of EEG. 2nd edition. Oxford University Press, New York
55. Pauluis Q, Baker SN, Olivier E (1999) Emergent oscillations in a realistic network: the role of inhibition and the effect of the spatiotemporal distribution of the input. J Comput Neurosci 6:27–48
56. Poe GR, Nitz DA, McNaughton BL, Barnes CA (2000) Experience-dependent phase-reversal of hippocampal neuron firing during REM sleep. Brain Res 855:176–180
57. Ray S, Maunsell JHR (2010) Differences in gamma frequencies across visual cortex restrict their possible use in computation. Neuron 67:885–896
58. Ray S, Maunsell JHR (2011) Network rhythms influence the relationship between spike-triggered local field potential and functional connectivity. J Neurosci 31:12674–12682
59. Remondes M, Schuman EM (2002) Direct cortical input modulates plasticity and spiking in CA1 representation along the proximodistal axis of CA1. Neuron 68:127–137
60. Salvador R, Silva S, Basser PJ, Miranda PC (2011) Determining which mechanisms lead to activation in the motor cortex: a modeling study of transcranial magnetic stimulation using realistic stimulus waveforms and sulcal geometry. Clin Neurophysiol 122:748–758

61. Schmidt R, Diba K, Leibold C, Schmitz D, Buzsáki G, Kempter R (2009) Single-trial phase precession in the hippocampus. J Neurosci 29:13232–13241
62. Schomburg EW, Anastassiou CA, Buzsáki G, Koch C (2012) The spiking component of oscillatory extracellular potentials in the rat hippocampus. J Neurosci 32:11798–11811
63. Schomburg EW, Fernández-Ruiz A, Berényi A, Mizuseki K, Anastassiou CA, Koch C, Buzsáki G (2014) Theta phase segregation of input-specific gamma patterns in entorhinal-hippocampal networks. Neuron 84:470–485
64. Singer W, Gray CM (1995) Visual feature integration and the temporal correlation hypothesis. Annu Rev Neurosci 18:555–586
65. Takahashi H, Magee JC (2009) Pathway interactions and synaptic plasticity in the dendritic tuft thalamus: link between the medial prefrontal cortex and the hippocampus. Brain Res Bull 71:601–609
66. Thielscher A, Opitz A, Windhoff M (2011) Impact of the gyral geometry on the electric field induced by transcranial magnetic stimulation. Neuroimage 54:234–243
67. Varga C, Golshani P, Soltesz I (2012) Frequency-invariant temporal ordering of interneuronal variation in visual cortex. J Neurosci 29:11540–11549
68. Vertes RP, Hoover WB, Szigeti-Buck K, Leranth C (2007) Nucleus reuniens of the midline voltage attenuation along CA1 pyramidal neuron dendrites. J Physiol 568:69–82
69. Wang C, Ulbert I, Schomer DL, Marinkovic K, Halgren E (2005) Responses of human anterior cingulate cortex microdomains to error detection, conflict monitoring, stimulus-response mapping, familiarity, and orienting. J Neurosci 25:604–613
70. Wu K, Canning KJ, Leung LS (1998) Functional interconnections between CA3 and the dentate gyrus revealed by current source density analysis. Hippocampus 8:217–230
71. Xing D, Yeh CI, Shapley RM (2009) Spatial spread of the local field potential and its laminar variation in visual cortex. J Neurosci 29:11540–11549
72. Yeckel MF, Berger TW (1990) Feedforward excitation of the hippocampus by afferents from the entorhinal cortex: redefinition of the role of the trisynaptic pathway. Proc Natl Acad Sci USA 87:5832–5836
73. Ylinen A, Bragin A, Nádasdy Z, Jandó G, Szabó I, Sik A et al (1995) Sharp wave-associated high-frequency oscillation (200 Hz) in the intact hippocampus: network and intracellular mechanisms. J Neurosci 15:30–46

FSC
www.fsc.org
MIX
Papier | Fördert
gute Waldnutzung
FSC® C083411

Kolibri 360 GmbH
Ferdinand-Jühlke-Straße 7
99095 Erfurt, Deutschland
produktsicherheit@kolibri360.de